Patrick Moore's Practical Astronomy

Other Titles in this Series

Navigating the Night Sky
How to Identify the Stars and Constellations
Guilherme de Almeida

Observing and Measuring Visual Double Stars
Bob Argyle (Ed.)

Observing Meteors, Comets, Supernovae and other transient Phenomena
Neil Bone

Human Vision and The Night Sky
How to Improve Your Observing Skills
Michael P. Borgia

How to Photograph the Moon and Planets with Your Digital Camera
Tony Buick

Practical Astrophotography
Jeffrey R. Charles

Pattern Asterisms
A New Way to Chart the Stars
John Chiravalle

Deep Sky Observing
The Astronomical Tourist
Steve R. Coe

Visual Astronomy in the Suburbs
A Guide to Spectacular Viewing
Antony Cooke

Visual Astronomy Under Dark Skies
A New Approach to Observing Deep Space
Antony Cooke

Real Astronomy with Small Telescopes
Step-by-Step Activities for Discovery
Michael K. Gainer

The Practical Astronomer's Deep-sky Companion
Jess K. Gilmour

Observing Variable Stars
Gerry A. Good

Observer's Guide to Stellar Evolution
The Birth, Life and Death of Stars
Mike Inglis

Field Guide to the Deep Sky Objects
Mike Inglis

Astronomy of the Milky Way
The Observer's Guide to the Southern/Northern Sky Parts 1 and 2
hardcover set
Mike Inglis

Astronomy of the Milky Way
Part 1: Observer's Guide to the Northern Sky
Mike Inglis

Astronomy of the Milky Way
Part 2: Observer's Guide to the Southern Sky
Mike Inglis

Observing Comets
Nick James and Gerald North

Telescopes and Techniques
An Introduction to Practical Astronomy
Chris Kitchin

Seeing Stars
The Night Sky Through Small Telescopes
Chris Kitchin and Robert W. Forrest

Photo-guide to the Constellations
A Self-Teaching Guide to Finding Your Way Around the Heavens
Chris Kitchin

Solar Observing Techniques
Chris Kitchin

How to Observe the Sun Safely
Lee Macdonald

The Sun in Eclipse
Sir Patrick Moore and Michael Maunder

Transit
When Planets Cross the Sun
Sir Patrick Moore and Michael Maunder

Light Pollution
Responses and Remedies
Bob Mizon

Astronomical Equipment for Amateurs
Martin Mobberley

The New Amateur Astronomer
Martin Mobberley

Lunar and Planetary Webcam User's Guide
Martin Mobberley

Choosing and Using a Schmidt-Cassegrain Telescope
A Guide to Commercial SCT's and Maksutovs
Rod Mollise

The Urban Astronomer's Guide
A Walking Tour of the Cosmos for City Sky Watchers
Rod Mollise

Astronomy with a Home Computer
Neale Monks

More Small Astronomical Observatories
Sir Patrick Moore (Ed.)

The Observer's Year
366 Nights in the Universe
Sir Patrick Moore (Ed.)

(continued after index)

A User's Guide to the Meade LXD55 and LXD75 Telescopes

Martin Peston

With 119 Figures

Springer

Martin Peston
Romford
Essex, UK
pestonm@yahoo.co.uk

British Library Cataloguing in Publication Data
A catalogue record for this book is available from the British Library

Library of Congress Control Number: 2006938890

Patrick Moore's Practical Astronomy Series ISSN 1617-7185
ISBN-10: 0-387-36489-7   e-ISBN-10: 0-387-68264-3
ISBN-13: 978-0-387-36489-6   e-ISBN-13: 978-0-387-68264-8

Printed on acid-free paper.

© 2007 Springer Science+Business Media, LLC
All rights reserved. This work may not be translated or copied in whole or in part without the written permission of the publisher (Springer Science+Business Media, LLC, 233 Spring Street, New York, NY 10013, USA), except for brief excerpts in connection with reviews or scholarly analysis. Use in connection with any form of information storage and retrieval, electronic adaptation, computer software, or by similar or dissimilar methodology now known or hereafter developed is forbidden.
The use in this publication of trade names, trademarks, service marks, and similar terms, even if they are not identified as such, is not to be taken as an expression of opinion as to whether or not they are subject to proprietary rights. Neither the publisher nor the author accepts any legal responsibility or liability for personal loss or injury caused, or alleged to have been caused, by any information or recommendation contained in this book.

Printed in the United States of America.

9 8 7 6 5 4 3 2 1

springer.com

Acknowledgements

Writing a book is not a trivial task. Lots of planning, hard work and many late nights have gone into producing it. I decided to write a book at probably the busiest time of my life, which included getting married and moving house. Somehow, I managed to find the time to sit down with a few ideas and 'put pen to paper', so to speak.

First and foremost, I would like to thank John Watson of Springer-UK, for his assistance in putting forward my proposal to Springer.

I would also like to thank Springer New York and Editors Harry Blom and Chris Coughlin, for their help and advice, and above all, their flexibility and understanding regarding timescales.

My sincerest gratitude to Alan Marriott who, without doubt spent tireless hours producing the illustrations notably, the numerous sky maps in the appendices and several drawings in the main chapters. Alan managed to produce excellent diagrams, despite my poor initial descriptions.

A very special thanks to Jack Martin and Steve Ringwood who spent many long hours proof reading my chapters. Their honest comments and advice were very much appreciated. Without their help and contribution, I would not have been able to complete the book on time.

I would also very much like to thank the following LXD owners for contributing images of the Moon, Planets and Deep Sky objects:

–David Kolb (Kansas, USA)
–George Tarsoudis (Alexandroupolis, Greece)
–Dieter Wolf (Munich, Germany)

Their images show what can be achieved with an LXD telescope and modest imaging equipment.

Sincere thanks go to Kevin Downing (Washington State, USA), for contributing images of his LXD75 N-6 Newtonian Reflector. Kevin went out of his way to send me the images, despite contacting him at very short notice.

Thanks to Meade USA contacts Chris Morrison and Scott Roberts for expressing their interest in the book, and their kind permission to use library images and literature regarding the LXD telescope series.

A big thankyou goes to Dave Lawrence of Telescope House UK. Dave put me in direct contact with Meade USA and was always happy to answer my LXD related queries. I would also like to thank Mike Cook for helping me acquire my very first LXD55 from Telescope House.

Many thanks to Don Gennetten (JMI Mobile), David Nagler (Tele Vue) and William Optics for their kind permission to use images from their accessories catalogs.

Acknowledgements

Also thanks to Richard Harris (LXD55.Com) and P Clay Sherrod (Arkansas Sky Observatory) for their kind permission to promote their telescope upgrade services. I would also like to thank them along with Mike Weasner (Mighty ETX, LXD websites) for their kind words of encouragement and permission to list their websites in the book.

Warm thanks go to the members of Loughton Astronomical Society UK for their continuous support and encouragement over the many years. An extra special thanks goes to Chris Williams for supplying images of his LXD SN-10 telescope and Steve Richards for his technical advice.

I would like to thank my friend John Servis who fortunately, manages a Cartridge World UK outlet. John supplied me with a consignment of ink and paper, which went towards the final drafts of the book before sending them off to the publisher. I would also like to thank my friend Jason Levy for formatting the Meade telescope images from Mac to PC.

Special heartfelt thanks go to my Mum and Dad. I wouldn't be where I am today if it wasn't for their encouragement and support. Ever since they purchased my first telescope, I have never looked back... now I look up! They always said that I should write an astronomy book! Well. Here it is!

And finally but by no means least, special thanks goes to my dear wife Ellayne for putting up with my constant 'I am working on the book!' comments and the very late nights. She put up with me the whole time I was locked away in the back room with the computer (or the 'other wife' as she calls it!). Ellayne has always been there for me. Her support and great sense of humor has kept me going, and I would never have achieved this without her. I therefore would like to dedicate this book both to my wife Ellayne and my Mum and Dad. I am eternally grateful.

This book has realized one of my lifelong ambitions and I thank everyone who has been involved.

It's been fun!

Martin Peston
London, UK
September, 2006

Author with AR-6 LXD55 Refractor mounted on an LXD75 mount.

Abbreviations

CCD .... Charge Coupled Device
DEC .... Declination
ECT .... Estimated Completion Time
GPS .... Global Positioning System
NCP .... North Celestial Pole
OTA .... Optical Tube Assembly
RA .... Right Ascension
SCP .... South Celestial Pole
SLR .... Single Lens Reflex

Contents

1. Introduction 1
   - The Goto Revolution — 2
     - To Goto or not to Goto? — 2
   - About This Book — 3

2. Astronomy as a Hobby 5
   - Introduction — 5
     - The Sky Above Us — 6
     - The Earth in Motion — 7
     - The Celestial Sphere — 8
     - Sky Coordinates — 9
   - Absolute Beginners — 11
     - Back to Basics — 11
     - Naked Eye Observing — 12
     - Finding Your Way Around the Sky — 12
     - Seeing Conditions — 18
     - Recording Your Observations — 19
     - Getting More Information — 20
   - Conclusion — 22

3. Choosing an LXD Telescope 23
   - Tools of the Trade – Introduction to Telescopes — 24
     - The Refractor — 24
     - The Newtonian Reflector — 26
     - Schmidt-type Telescopes — 27
     - Telescope Tube Essentials — 29
     - Telescope Mounts — 31
     - Telescope Stands and Tripods — 34
   - Personal Factors to Consider when Choosing a Telescope — 38
     - Purpose — 38
     - Quality and Cost — 40
     - Portability — 40

The LXD Telescope Series	40
AR-5 and AR-6 Refractors	41
N-6 Newtonian Reflector	41
SN-6, SN-8, SN-10 Schmidt-Newtonian Telescopes	44
SC-8 Schmidt-Cassegrain Telescope	44
Features of the LXD Series	47
Mount Features	48
Conclusion	52

4. Setting up the Telescope 55

Setup Tasks	56
Assembling the Telescope (ECT: 2 to 3 Hours)	56
Handling the Telescope	57
Balancing the Telescope	60
Polar Viewfinder Alignment	63
Polar Viewfinder Shaft Alignment Test (ECT: 10 Minutes)	64
Aligning the Polar Viewfinder With the RA Shaft	65
The Polar Home Position (ECT: 5 minutes)	67
Aligning the OTA With the RA and Dec axes	68
The LXD Motor Drives	72
Autostar Backlash Compensation Utility (ECT: 15 to 25 Minutes)	74
Physical Motor Alignment (ECT: 45 to 60 Minutes)	74
Tripod Setup (ECT: 5 to 10 Minutes)	80
Finderscope Alignment (ECT: 15 to 30 Minutes)	82
Focuser Adjustments (ECT: 10 to 15 Minutes)	84
Conclusion	85

5. Polar Alignment and Goto Setup 87

Using Polaris to Find the North Celestial Pole	87
Determining Latitude	89
Aligning the RA Shaft with the North Celestial Pole	89
Tripod Setup	90
Latitude Adjustments	91
Locating Polaris in the Polar Viewfinder	91
Traditional Methods for Polar Aligning	94
Setting up the Goto Facility	98
Easy Align (ECT: 5 Minutes)	99
One Star Alignment (ECT: 5 to 7 Minutes)	99
Two Star Alignment (ECT: 5 to 6 Minutes)	100
Three Star Alignment (ECT: 6 to 7 Minutes)	100
Alignment Setup – Success or Failure?	101

Southern Hemisphere Alignment	102
Non-Goto Operation of the Telescope	102
Conclusion	103

6. First Night's Observing 105

An Experience to Forget...	105
Location! Location! Where to Observe	106
Equipment! What to Take Out	107
Telescope Setup	108
Slewing the Telescope	109
End of the Night – Packing Up	111
Packing Up in Public Locations	111
Parking the Telescope	111
Sleep Scope	112
Summary	113

7. Telescope Operations, Abilities and Observing Techniques.. 115

Observing in Comfort	115
Observing Celestial Objects Across the Meridian	116
How Faint Can You See – Limiting Magnitude	117
How Much Detail Can You See – Resolving Power	118
How Far Can You Zoom – Magnification Power	118
How Much Light from an Object Can You See – The Exit Pupil	119
Observing Techniques	119
Dark Adaptation	119
Averted Vision	120
Telescope Operations	120
Tracking Objects	121
Using the Telescope Without Goto to Find Objects	123
Using Goto to Find Objects	124

8. The Universe at a Touch of a Button 129

Introduction	129
Guided Tour	129
The Autostar Object Database	130
The Solar System	130
Constellations	145
Stars	145
Deep Sky	147
Using Autostar to Identify Objects	153
Browsing the Autostar Object Database	153

User Object	155
Suggest	155
Summary	155

9. Connecting to a Personal Computer ... 157

Introduction	157
Connecting the Autostar to a PC	157
Computer Setup	159
Upgrading the Firmware	160
Other Features of the ASU	162
Autostar Cloning	162
Connecting Autostar to Other Devices	163

10. Taking Images ... 165

Traditional Astrophotography	165
Choosing the Right Equipment	165
Choosing the Right Camera Film	166
Astrophotography with an LXD Telescope	167
Digital Imaging	172
Webcam Imaging	173
CCD Imaging	174
Digital Cameras	175
Video Cameras	176
Focusing Tips	177
Image Processing	177
Conclusion	177

11. Keeping Your Telescope in Peak Condition ... 179

Collimating Your LXD Telescope for Pin-Sharp Images	179
Testing the Collimation of a Telescope	180
Collimating the AR Refractor (ECT: 1 to 2 Hours)	182
Collimating the SNT (ECT: 1 to 2 Hours)	183
Collimating the SCT (ECT: 1 to 2 Hours)	185
Collimating the N-6 Reflector	187
Collimation Summary	187
Cleaning the Telescope	189
Cleaning the Optics	189
General Maintenance	191
The Ultimate Telescope Tune-up!	192
Storing the Telescope	192
Indoor Storage	193

Contents

Storing the Telescope Outdoors	193
Summary	194

12. Gadgets and Gizmos 195

Introduction	195
Eyepieces	195
Erecting (Terrestrial) Eyepieces and Prisms	197
Barlow Lens	197
Focal Reducer	198
Star Diagonals	198
Finderscopes	199
Filters	199
Filter Wheels	202
Binocular Viewers	203
Tripods	203
Motorized Focusers	203
GPS Add-Ons	203
Bluetooth Connectivity	204
Power Supplies	204
Dew Shields and Heaters	205
Meade #909 Accessory Port Module (APM)	205
Piggyback Camera Mounts	206
LXD Carry Handle	206
Autostar Placeholder	207
Telescope Covers and Cases	207
Summary	208

13. Where Did It All Go Wrong? 209

Alignment Stars are not in the FOV During Goto Setup Procedure – Alignment Failure is Displayed on the Autostar	210
Objects not Found in the FOV after a Goto is Performed	210
The RA or Dec Axes Do not Turn When Autostar Arrows Keys are Pressed	211
Autostar Displays 'Motor Unit Failure'	211
Poor Tracking and Backlash Problems	211
Text on the Autostar Screen Appears Blurry and Slow to Display	212
Autostar Unexpectedly Reboots	212
The Arrow Markers Located on the Side of Each Axis are not Aligned in Polar Home Position	213
Summary	214

Appendix A: Lists and Charts of Autostar Named Stars 215

Appendix B: Object Lists 229
   B1. Autostar Constellation List   229
   B2. Autostar Messier Objects List   231
   B3. Autostar Caldwell Objects List   234
   B4. Annual Meteor Showers   237

Appendix C: Autostar Menu Options 239

Appendix D: References and Further Reading 241
   D.1 List of Useful Websites   241
      LXD Telescope Websites   241
      Other Websites and Resources   243

Appendix E: Astronomical Image Information 247
   E.1 David Kolb Images   247
   E.2 George Tarsoudis Images   248
   E.3 Dieter Wolf Images   249
   E.4 Author's Images   250

Index 251

CHAPTER ONE

Introduction

I have owned telescopes for over 25 years since I was a young lad. I purchased an LXD55 AR-6 Refractor in 2002, and was one of the first to own one in the UK. I am also a proud owner of an LXD75 SC-8. Armed with these two very different telescopes, I have spent many hours searching the skies for interesting objects using Meade's Autostar Goto facility.

My motivation to write a book about the LXD Goto telescope series, first came from comments about an LXD55 AR-6 Refractor review, that was published on the LXD55.com website. From then on, I have had regular emails from people asking technical questions about the telescope, and which model is best suited for them. Whilst attending Star parties in the UK, I found that many LXD owners would struggle to use them even at a basic level, especially if they have never owned or used an equatorially mounted Goto telescope before. Since the first LXD55 models came out in early 2002, owners have struggled to find useful information to help them use the telescopes to the best advantage. There have been mixed reactions about its quality and performance. Hence, this book is directed towards those who are new to Goto and the LXD telescope.

So why choose an LXD telescope? Many telescopes with Goto capabilities currently on the market, tend to be alt-azimuth mounted in design. They use either a single arm, or a fork mount to control the movement of the telescope tube. The LXD series however, uses a standard German Equatorial Mount (GEM) design, retro-fitted with Goto capability. Even though setting up a GEM with Goto is more involved than an Alt-azimuth mount, the GEM-Goto design of the LXD telescope is what has appealed to astronomers, both beginners and amateurs alike.

The Goto Revolution

Fifteen to twenty years ago if you wanted a telescope which could automatically find celestial objects in the night sky at a touch of a button, you would only be able to find them in custom-built observatories, or in professional telescopes on top of remote mountains.

The introduction of the Goto telescope to the astronomical community was heralded as a breakthrough in amateur astronomical observing.

Meade introduced their Autostar Goto handset back in 1999. They were supplied with the ultra portable popular ETX90EC. Since then, newer models of the Autostar handset have been released, and are supplied as standard with many Meade telescopes, including the ETX, LX90 and of course, the LXD telescope series. More information about Autostar in Chapter 3.

With today's skies seriously affected by light pollution, Goto has become *the* observing solution for astronomers to assist them in searching for those faint fuzzy objects.

To Goto or not to Goto?

Owning a Goto telescope has changed my personal attitude towards observing. I used to spend hours trying to search for elusive objects that were difficult to locate, and a whole night's observing would pass by with only a few distinctive objects seen. Of course, I used to enjoy those observing sessions but nowadays, I know that with my Goto telescope set up correctly, I can find many objects in a single observing session. I am not racing to see how many objects I can find in one night, it is the case of confidently knowing that an object I selected from the Autostar handset will be visible in the eyepiece.

I consider myself an experienced observer, and feel that using a Goto telescope has enhanced my observing skills rather than depreciate them, which is what many people are led to believe when they consider using these types of telescopes. If I was a beginner, new to astronomy and had the opportunity to purchase a telescope with Goto capabilities, I would expect the telescope to do all the complicated setting up procedures for me, and successfully point to an object selected from the Goto handset. Does this make me a lazy astronomer? Will I forget how to star hop? Or will I become completely dependent upon the electronics of the telescope to know my way around the sky?

Whether a beginner should buy a Goto telescope for their first observing instrument is a subject of many discussions. My personal opinion on the questions asked above? Well, I firmly believe that for someone who has very little astronomical observational experience, should learn the night sky with nothing more than the naked eye, or a good pair of binoculars. However, for someone who has some experience of using a non-Goto telescope, but would like to upgrade their telescope to one with advance electronic features, then a Goto telescope is ideally suited for them.

If you happen to own an LXD telescope and you don't know the first thing about using it, then I hope this book will be of use to you. I hope to have provided you with

Introduction

all the necessary information to help you get to grips with Goto. Unfortunately, the manuals supplied with these telescopes only go some way to describe the real practical aspects of using a Goto telescope. This is why some sections in this book notably, Chapter 4 (Telescope Setup) elaborates upon the instruction manual. I would still suggest however, that you browse through the manual, to fully appreciate the setup procedures described in this book.

About This Book

There is a deliberate chapter order to this book; assembling, setting up, and using the telescope.

- Chapter 2 provides you with a brief guide to the celestial sphere and jargon used elsewhere in the book.
- Chapter 3 explains the types of telescopes that are on offer and what LXD model is best suited for you.
- Chapter 4 describes how to set up the telescope prior to observing.
- Chapter 5 helps you to Polar align.
- Chapter 6 explains what happens during a typical night's observing with a Goto telescope.
- Chapter 7 provides a mixture of useful information, such as the operations and abilities of the telescope, Autostar features and general observing techniques.
- Chapter 8 is a complete breakdown of the Autostar Object menu including images taken by owners of the telescope.
- Chapter 9 describes how to connect to a personal computer to control the telescope remotely and upgrade the Autostar internal firmware.
- Chapter 10 explains the equipment required to take images through the LXD telescope.
- Chapter 11 shows how to look after your LXD telescope.
- Chapter 12 provides information about the accessories that are available for the LXD telescope series.
- Chapter 13 is the troubleshooting chapters and provides solutions to some of the most common problems encountered by LXD owners.

 A quick mention about the term 'LXD' used throughout the book. When the term 'LXD' is used without the suffix '55' or '75', it represents both models in the telescope series. For descriptions that apply to specific models only, then the terms 'LXD55' and 'LXD75' will be used.

 I have provided a detailed set of Constellation star charts, depicting the location of every 'Named Star' listed in the Autostar database (Appendix A). These charts will be essential during the Goto setup procedures outlined in Chapter 4.

 Appendix B at the back of the book contains useful information, such as lists of deep sky objects and constellations.

There is an Autostar navigation table in Appendix C, which will help you find your way around Autostar's extensive menu systems. There were several Autostar firmware releases from Meade, and some menu options have been relocated or added during the drafting of the book, so the menu lists may change in subsequent firmware updates. The firmware I originally used for menu descriptions was 32Eh. I have now included changes up to version 42Ed. The Index also contains a full list of Autostar options and where they are described in the book.

I have decided not to delve too deeply into various topics such as mechanical modifications or detailed programming of the computer handset, as they are covered elsewhere in other books or Internet resources.

I hope to think that this book will inspire further research about the LXD telescope series, considering the amount of information that is now freely available on the Internet. Hence, lists of useful websites and further reading are provided in Appendix D. And finally, Appendix E provides details about the astronomical images that were used in the book.

CHAPTER TWO

Astronomy as a Hobby

Introduction

It is said that Astronomy is the oldest of the sciences. For thousands of years civilisations have looked up at the sky, the Sun, the Moon and the Planets and tried to link their destinies to them. Huge stone observatories were built to predict the movements of the Sun, Moon and the Stars.

Over the past few centuries the scientific pursuit of knowledge has driven mankind to explore the heavens, in an ever-increasing depth in order to understand the Universe around us. New theories and technical advances have expanded mankind's knowledge of the cosmos, to such an extent, that we can estimate the start of the Big Bang to within millionths of a second, or estimate what the Universe might look like billions of years into the future.

Astronomy is no longer confined to the Royal Court of Kings and Queens or forms part of a wealthy gentleman's leisurely pursuit. Nowadays, to be an astronomer you don't have to build telescopes or grind mirrors, draw your own star maps or manually calculate the next phase of the moon or position of a planet. A lot of astronomers still like to do these tasks, including myself, and they are worthwhile pursuits for the more dedicated. But for the beginner it could sound daunting that to become an amateur astronomer you have to be an engineer, mathematician, physicist or all three rolled into one. This is not the case and you don't have to be experienced in a scientific discipline to enjoy astronomy as a hobby. The hobby of astronomy is not a 'black art' as most people make it out to be.

Everyone has their own reasons for pursuing astronomy as a hobby. Some like the thrill of one day discovering something such as an Asteroid, Comet or exploding Star.

Others find it a sociable hobby in order to meet with like-minded people to talk about life, the Universe and everything!!

The hobby is becoming more popular amongst the general public. One of the reasons for it being more popular nowadays is the use of equipment that makes it easier to find objects in the sky. Thanks to innovative technology introduced into the telescope market over the past decade, thousands of astronomical objects are now within reach of the amateur astronomer.

'Instant astronomy' is here. You can now buy telescopes at all levels, from the simple point-and-observe to all-seeing all-dancing Goto systems with GPS Satellite tracking that even tells you what to see when you are out observing. The only thing they cannot do is keep you warm during those cold dark winter nights! Even today's professional astronomers never physically look through large telescopes they are using. It's all done remotely through computer control. A typical amateur astronomer's setup nowadays is almost at par with the professionals' of ten or more years ago.

The popular misconception that most people have when buying a telescope with Goto features in particular, is that they do not need any prior knowledge of the night sky. The telescope will do everything for them and they will learn the sky through the use of the telescope features. Unfortunately this is not always the case and it always invariably leaves the user frustrated and disappointed after a failed night's attempted observing.

While it is true that once you get to grips with using the telescope correctly, you can find all manner of astronomical objects in the night sky, learning to use the telescope from scratch; switch on and off you go, is more difficult than originally realised. The use of a Goto telescope still requires a basic knowledge of the night sky.

The rest of this chapter is dedicated to explaining some of the very basic concepts in astronomy. It is primarily directed to those who have very little experience, or no knowledge of the night sky. Those who have a grasp of the basic concepts can skip the next few sections if they so wish.

The Sky Above Us

Most people take the sky above them for granted. This may sound an odd thing to say, but it's often true. Only during something exceptional, which is normally publicised through the national media, do people then take notice that the sky is displaying something worth looking at, such as a Solar or Lunar Eclipse, Meteor Shower or usual lighting conditions.

Unfortunately the skies in the United Kingdom from the major cities are heavily light polluted, drowning out all but the Moon, some Planets and the brightest stars. The United States and other countries around the world no doubt also have their own light pollution problems in densely populated areas. There have been mixed responses by Governments to try and reduce the light pollution in the most populated urban areas with improved lighting. For now though, the only real way to fully appreciate the night sky in all its glory is to observe from a location a distance away from the cities and towns. However, most people make do with the light pollution situation and enjoy astronomy no less.

Astronomy as a Hobby

So, if you look up into the sky over a particular period of time what do you see?

If you happen to be fortunate to live in a place where there are dark skies and little light pollution, you will be able to see the Milky Way seen as a band of stars, arching the sky from horizon to horizon. I will discuss more about the Milky Way later in this chapter.

Over the period of a year you will see different constellations in the night sky on display throughout the seasons. Intermixed with these constellations you will see what appear to be slowly moving bright stars. These are in fact the Planets in our Solar System which course independently along the ecliptic, as they orbit the Sun through the Zodiacal constellations. Over the period of a month, you will see the phases of the Moon, and over the period of a day you will see the daily motion of the Sun.

The Earth in Motion

The changing displays of the constellations in the sky over a period of time are governed by the Earth's rotation (daily changes) and Earth's motion around the Sun (monthly and yearly changes).

We base our lives on the 24 hours day–night cycle. This is known as the Solar day, and is the length of time it takes for the Earth to spin on its axis once with respect to the Sun.

However, in relation to the Stars, the Earth spins on its axis once every 23 hours 56 minutes and 4 seconds. It is shorter than a solar day by about 4 minutes. This is the Earth's true period and is called the Sidereal day.

The reason for the difference in solar and sidereal days is that the Earth moves a specific distance along its orbit around the Sun in a single day. The position of the constellations which, for all intense purposes, are fixed in their positions in the sky will be seen by an observer on the Earth to have moved a small distance in the sky (about $1°$), with respect to the Earth's position in its orbit around the Sun.

The difference of about four minutes between the solar day and sidereal day means that the constellation you see at a particular time on any given day will be in the same position in the sky, but 4 minutes earlier each night.

For example, a constellation that is due south at 10 p.m. at the beginning of a month, will be due south at 8 p.m. by the end of the month, almost 2 hours earlier. Hence over a period of 6 months, it will be due south some 12 hours earlier, i.e. during the day. In a single standard year, that constellation will be due south once again but 24 hours, one Solar day earlier. This doesn't mean however, that over longer time periods, such as decades or centuries, the winter constellations would be seen during the summer months and the summer constellations during the winter months, because of this one-day-per-year time drift; it's just that you can 'fit' $366\frac{1}{4}$ Sidereal days into one Solar year.

In other words, by the time the Earth has travelled round to the same point in the orbit again with respect to the Sun (one Solar Year), $365\frac{1}{4}$ Solar days or $366\frac{1}{4}$ Sidereal days will have transpired. The 'odd' quarter day is accounted for by having a 'Leap' day every 4 years.

The Celestial Sphere

The sky can be thought of a sphere with the stars projected onto it. Its coordinates are set horizontally and vertically, much like the longitude and latitude that we adopt for fixed positions on the Earth. In addition, the Earth's poles and equator are projected onto the sphere.

The Earth has its polar axis inclined at an angle of $23\frac{1}{2}°$ to the plane of its orbit around the Sun. This inclination gives rise to the seasons.

At different times of the year there are different constellations on display. This depends upon the Earth's position around its orbit of the Sun. As you can see from Figure 2.1, at different times of the year, the Earth is placed such that the night-time hemisphere is facing a particular group of stars. During the day the Sun is visible and because of its sheer brightness drowns out the other constellations that are in the sky

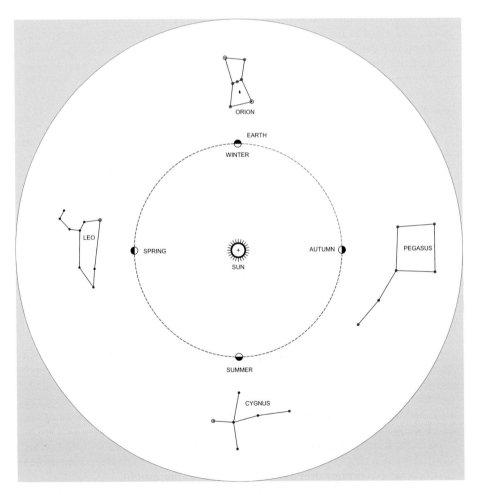

Figure 2.1. Earth's motion around the sun. Image courtesy of Alan Marriott.

Astronomy as a Hobby

at the time. If the Sun wasn't there, you would see a different set of stars during the daytime. As the Earth travels round the Sun in its orbit, the view of the stars from the Earth gradually changes from season to season.

The Earth's inclination to the plane of its orbit produces some interesting effects depending upon where you are standing on the surface of the Earth. If you happen to be at the North or South poles, the stars will move around the poles in a circular motion directly overhead to your position. The stars near the horizon will never rise or set and will appear to follow it all the way around in parallel paths during the course of the night. These are known as circumpolar stars.

If you are standing on the equator however, you will notice that all the stars will rise and set in huge semi-circles from East to West. None will stay above the horizon all night.

Most of us however, do not live on the equator or at the poles and so what we see is a star that rises, reaches its highest point (culminates) and then eventually sets, describing a great arch in the sky. Directly above an observer's head at the highest point of the sky is called the Zenith. Stars near the local Zenith do not stay above the horizon all night in most locations on the Earth, except for locations that are very close to the North and South poles (Figure 2.2).

Sky Coordinates

Astronomers use Sidereal time to work out which astronomical objects are visible in the sky at any given time. A Sidereal day is split into twenty-four 1 hour segments just like the normal clock time that we use. Using the Sidereal day makes it rather more convenient for astronomers than trying to split up 23 hours 56 minutes of the solar day with respect to the stars motion.

Stars rise, culminate and set at the same sidereal time every sidereal day. Hence, the stars are more or less fixed to the sidereal positions. This is useful if astronomers want to map the position of the stars in the sky. Stars have *proper* motions in which they slowly change positions in the sky over long periods of time as they orbit the galaxy. This motion, however takes hundreds, if not thousands of years and therefore their sidereal positions are not adversely affected in the immediate short term. Star Maps however are periodically updated every 50 or so years so that they can provide up to date positions of the stars.

Astronomers call the sidereal time of any given astronomical object, its Right Ascension (RA). The initial starting point is 0 hour 0 minute 0 second, which is known as the 'First point of Aries' and is incremented in hourly points eastwards from the initial start point. It is similar in concept to Greenwich Mean Time, which starts from the meridian at Greenwich, London where all other references to the Earth's Longitude are based upon. The right ascension is projected onto the celestial sphere along the celestial equator.

Note: the term 'First point of Aries' was first coined by the Greeks over 2000 years ago. Since then, this 'point' has moved into the constellation of Pisces, due to the Earth's polar axis drifting over a long period of time (precession).

Each 1 hour of RA projects 1/24th of a complete circle. Since a circle is 360° each 1 hour part represents 15°. So a star at 6 hours RA is 90° from the first point of Aries,

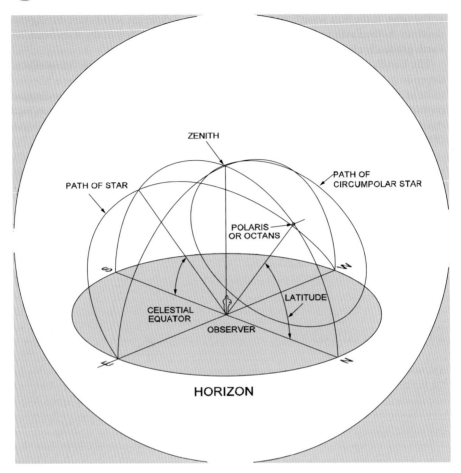

Figure 2.2. Observer's perspective of the celestial sphere. Image courtesy of Alan Marriott.

a star at 12 hours is 180° from the first point of Libra on the opposite side of the sky and so on.

Vertical coordinates of the celestial sphere are known as the declination (Dec) which are measured in angular intervals, degrees denoted by the degree symbol °, minutes denoted by the single quote symbol ′ and seconds denoted by the double quote symbol ″. These pairs of RA and Dec coordinates are often referred to as Equatorial or Celestial coordinates.

The declination of a star is also fixed, just as the right ascension so astronomers can pinpoint astronomical objects in the sky by using both their RA and Dec coordinates. The declination of an astronomical object is the angular distance from the celestial equator. It has positive coordinates up to 90° above the equator and negative coordinates below the equator down to minus 90°. Hence, if you see a star with a negative declination you will automatically correctly conclude that the star is located below the celestial equator.

For example, the brightest star in the northern hemisphere is Sirius whose coordinates are RA. 06 hours 45 minutes and Dec. $-16°\ 45'$. So the star is situated about $16°$ below the celestial equator.

Polaris however, is RA. 02 hours 31 minutes and Dec $+89°\ 16'$, which places it very close to the celestial pole. We shall be talking more about Polaris in Chapter 5.

In addition to the celestial equator, we have the Ecliptic. This is the apparent path that the Sun takes over the period of a year and represents the plane of the Earth's orbit. It makes an angle of $23\frac{1}{2}°$ to the celestial equator (Figure 2.1). Our Moon, the rest of the Planets and other bodies of the Solar System are located on or around the area of the ecliptic plane. The constellations that the ecliptic passes through are known as the Zodiacal constellations. Astrology attempts to draw upon some significance of the positions of the Planets and the Zodiacal constellations at specific times of the year, and how they are supposed to govern our daily lives.

The ecliptic and the celestial equator crosses at two points on the sphere. The first one is at the 'First point of Aries', which happens to be where RA measurement is started from. The second position is directly at $180°$ on the opposite side of the sky, at the position known as the 'First point of Libra'. When the Sun crosses these points we have the Equinoxes. On or around March 21st each year, the Vernal equinox occurs when the Sun moves northwards to above celestial equator from below. The Autumnal equinox occurs on or around September 23rd each year, when the Sun moves southwards to below the celestial equator.

I have discussed just a few of the main concepts of the celestial sphere in this section. The more advanced concepts are not required here as you only need to understand the basic concepts in order for you to use a Goto telescope.

Absolute Beginners

If you are just starting out in astronomy and have very little or no experience of the night sky, then I would not recommend that you jump straight into the deep end and start using a telescope immediately. It is best to start observing with nothing more than your eyeballs and a good set of sky maps. With a little dedication and patience, you will find observing a rewarding and enjoyable experience.

Back to Basics

The first steps in familiarising yourself with the sky is to go out at night, armed with nothing more than a red-light torch and a sky map. There are many excellent sky maps available. A sky map will assist you to recognise well known bright stars and constellations, and when they are visible at specific times of the year.

Even this simple task of finding well known stars and constellations to an absolute beginner is easier said than done. Simply going straight out and trying to match constellations with what you see on the sky map, can still be a difficult task, especially if you happen to live in an area where the skies are light polluted. The stars on the sky map might not depict what you can actually see in the sky. It is the locations of the brightest stars along with their proper names that you should familiarise yourself with. Once you know the positions of the well-known stars in the night sky, you will be able to set up telescopes that have Goto features.

Before you can begin use a sky map at night, what you need to do first is to get your bearings. You need to work out which way the Cardinal points; North, South, East and West are facing, as this will help you find your way around the sky. The steps below will help you determine the cardinal points at your location.

1. During daylight hours, take with you to the observing area, a compass or anything other means that will allow you to determine which way you are facing. For example, by electronic means such as a GPS.
2. Survey the observing area and find North using the compass. Make a mental note which way you are facing. For example, North could be facing towards your residence or towards a recognizable terrestrial landmark, such as a tree or distant hills.
3. Do the same for the other South, East and West cardinal points.
4. You may want to make a mark on a nearby object that represents a cardinal point so at night you will be able to determine the direction you are facing.

This basic procedure will help you not only find astronomical objects in the night sky, but will also help you towards setting up a telescope when the time comes. Now you are ready to find your way around the sky.

Naked Eye Observing

The Mark I eyeball is the best tool for you to use to help you find your way around the night sky. The eye is a remarkable astronomical instrument and has been used for millennia to observe the heavens long before optical aid was conceived.

Finding Your Way Around the Sky

Preparation

When you go out to look at the night sky, make sure you are prepared.

- Take a Torch with you - preferably one with a red light. This is essential in finding your way around the observing site, also for security and safety. This will allow you to dark adapt much easier, as bright white light tends to dazzle your eyes, making it more difficult for you to see faint astronomical objects. It takes approximately 30 minutes or so for the eye to dark adapt. More on dark adaption techniques later in Chapter 7.

Astronomy as a Hobby

- Use a notepad and pencil to record your observations. A pencil is preferable to an ink pen as the ink tends to dry out under cold conditions.
- Carry a time-keeping device with you such as a watch, so that you can make timed observations.
- Dress up warmly. Even during the summer nights, conditions can be deceptively cold. The head should be kept covered as it is where the body heat is lost the most.
- Take something to sit on. Being comfortable contributes towards your concentration. So find somewhere at the observing site where you will be able to take rest periods.
- If you are observing from a remote location remember to take something to eat and a hot drink.
- Finally, take a sky map to help you navigate your way around the sky.

Armed with this basic information you are now prepared to discover the wonders of the night sky.

Sky Maps

Sky maps aid the astronomer in finding objects in the night sky. Even with Goto systems, sky maps are still very much used today to help track down the most interesting objects in the sky.

A typical set of sky maps consist of the following.

- Maps showing Northern and Southern Celestial Pole regions.
- Maps depicting constellations per season or, Maps showing segments of right ascension and declination. For example, Map 1 RA. 0 to 6 hours Dec0-90°, Map 2 RA. 6 to 12 hours and so on.
- An index included with each map highlighting objects of interest such as double stars, nebulae and galaxies.
- A depiction of the Milky Way spread across the maps in appropriate places.

Some sky maps show the northern or southern hemispheres of all of the constellations all on one map. They are probably not as useful as maps which show partial segments of the sky, as they do not show a realistic view, seen by an observer who is standing on the ground looking up. Planispheres are commonly used as well to determine what is currently visible in the night sky.

Astronomy magazines and computer programs also provide sky maps. They show events which are currently happening in the sky, such as the positions of the Moon and Planets, as well as information about Comets and Asteroids.

Recognizing the Constellations

When you go outside into your garden armed with your sky map, where do you start?

If you carried out the exercise in the previous section for getting your bearings, North, South and so on... then you should be able to match the sky map with the direction you are facing.

There are 88 designated constellations in the sky. A good number of these known constellations have distinct patterns which are unmistakable. These recognisable

A User's Guide to the Meade LXD55 and LXD75 Telescopes

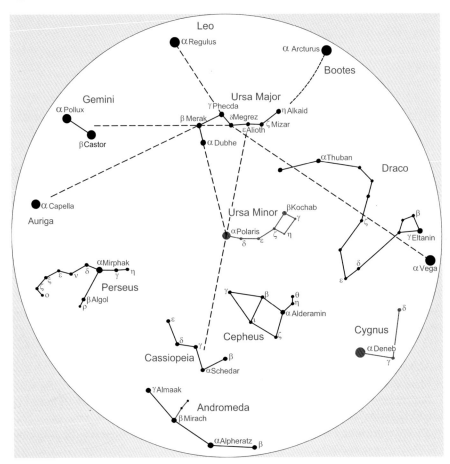

Figure 2.3. Using the big dipper to find other constellations.
Image courtesy of Alan Marriott.

constellations in the sky can be used to identify bright stars and other less distinct patterns surrounding them. A list of the constellations is in Appendix B.

For those living in the northern hemisphere, the best constellation to start with is the Great Bear near the North Polar area. The Great Bear has an unmistakable pattern. Its seven stars form the shape of a saucepan, farmers plough or Big Dipper. This constellation can be used to find several other constellations nearby and more importantly, the north celestial pole. Telescopes that track the stars require alignment to the north celestial pole. I will describe how to find the north celestial pole and align a telescope to it in Chapter 5.

Rotate Figure 2.3 so that it matches the orientation of the Big Dipper as you see it in the sky and follow the imaginary lines shown to the relevant constellation. As you can see the two pointer stars Merak and Dubhe point towards the North Pole Star Polaris. The bowl of the Big Dipper faces towards the constellation Leo. Follow the

Astronomy as a Hobby

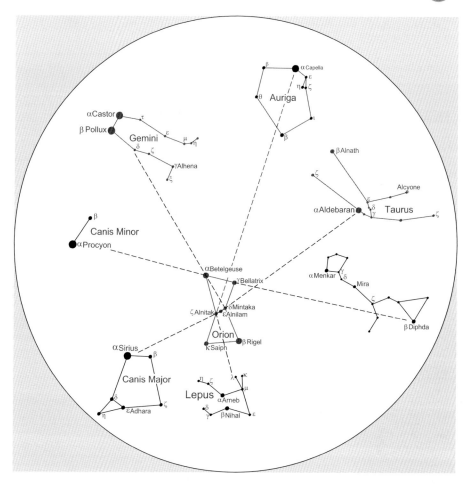

Figure 2.4. Using Orion to find other constellations. Image courtesy of Alan Marriott.

arch of stars down from the bowl of the Dipper to find the bright star Arcturus in the constellation Boötes.

Another prominent constellation used to find other nearby constellations is Orion the Hunter in the winter skies.

The most prominent area of Orion is three bright stars Alnitak Alnilam and Mintaka which line up diagonally.

Follow the stars diagonally upwards and you will come to the bright star Aldebaran in Taurus. Keep going in the same direction as the diagonal line and you will eventually come to a small group of stars called the Pleiades or Seven Sisters. Travel diagonally downwards from the three stars in Orion and you will come to the brightest star in the northern sky, Sirius in the constellation of Canis Major. Other nearby constellations are located by drawing imaginary paths through various combinations of stars in the Orion constellation (Figure 2.4).

As you can see, this method of star hopping is very powerful for navigating your way around the sky.

Orion and the Big Dipper are just two of many prominent recognisable constellations in the sky. In no time you will be able to find your way around the sky with confidence and amaze your colleagues, friends and family at the ease of stargazing.

The Milky Way

Our home – the Earth and the Solar System is a member of a spiral galaxy which contains about 200 billion stars. We live within one of the arms in the plane of the galaxy. Hence, we see our galaxies spiral arm as a faint band of light arching the sky rather than a familiar spiral shape like we see with other galaxies external to our own. Most sky maps represent the Milky Way as a graphical illustration using a different colour to that of the map background.

Unfortunately, the Milky Way is not often seen in or near large cities as light pollution tends to brighten the sky, such that it drowns it out. Those who are fortunate enough to see it find it to be a beautiful and awe inspiring sight. In very dark skies even with the naked eye you can see faint structure and dark rifts (gaps) in certain regions. Almost all of the objects we see in the night sky with the naked eye are part of the Milky Way – the one exception is the Andromeda Galaxy. Through binoculars or telescopes the Milky Way is resolved into millions of stars and you can observe these patches even in skies with minimum light pollution, if you know where to look of course!

The Brightness of Stars

Apparent or Relative Magnitude. Magnitude is the term given to the brightness of a star. The apparent magnitude of a star is the term used to describe how visually bright a star is to the naked eye, regardless of how far away it is from us. You could have a large star very far away or a smaller star close by and they can be similar in magnitude.

It is not just stars that have apparent magnitudes. All celestial objects in the sky such as the Sun, Moon, Planets, Nebulae and Galaxies have their brightness determined the same way. These objects however, are not pin-points of light when seen through binoculars or telescopes, but have a real apparent size in the sky. They are known as extended objects. Hence, astronomers also measure the apparent *surface brightness* of certain extended objects which in most cases are fainter than the apparent magnitude values assigned to them. The magnitude is determined as if all of the light of these extended objects are combined into a single point.

The measure of a stars' brightness was first determined by Hipparchus in the second century BC and later classified in Ptolemy's star catalogue. The magnitude scale for objects seen with the naked eye was originally classified into six grades. First magnitude stars were the brightest and sixth magnitude stars the faintest. Nowadays, the magnitude scale is still based on the original concept but has been expanded to include both positive (+) and negative (−) values. The brightness of objects decreases as we go from negative to positive values along the scale. Table 2.1 shows the magnitude for various celestial objects.

Various methods of measurements show that the difference in brightness between two celestial objects of one magnitude is approximately 2.5 times. For example, a magnitude +2.0 star is 2.5 times as bright as a magnitude +3.0 star. Hence, the

Astronomy as a Hobby

Table 2.1. Magnitudes of Various Celestial Objects

Celestial Object	Magnitude
Sun	−26.7
Moon (full)	−12.7
Venus (at max brightness)	−4.4
Jupiter (at max brightness)	−2.4
Sirius	−1.46
Vega	+0.04
M31 Andromeda Galaxy	+4.8
North American Nebula	+9.0
Central Star of Ring Nebula	+16
Hubble Space Telescope faintest object	Approx +28

difference in brightness between a magnitude +1.0 star and a magnitude +6.0 star is almost 100 times.

The naked eye has the ability to visually determine an approximate magnitude of a star. In fact, a trained observer can determine the brightness to a tenth of a magnitude by comparing with other stars whose magnitudes are already known.

The dark-adapted naked eye can see faint stars under good conditions to about sixth magnitude in non-light polluted skies. This is known as the limiting magnitude. Optical aid such as binoculars or telescopes can see much fainter and depends upon the aperture of the instrument. I will talk about limiting magnitude of various instruments in Chapter 7. CCD imagers used in conjunction with telescopes can delve even deeper. The Hubble Space Telescope has a limiting magnitude of around +28.

You will find that all star maps provide graphical representations of magnitudes of stars. Black dots or circles of various sizes represent different magnitudes. The majority of star maps normally go down to magnitude +6, but the more specialised stars maps, such as Uranometria display stars that are much fainter.

Absolute Magnitude. So far I have talked about relative magnitude, where the magnitude values are determined by comparison with other similar stars. However, astronomers sometimes refer to a star as having a particular absolute magnitude.

A star's absolute magnitude is the apparent magnitude it would have if it was at a fixed distance away. The distance is 10 parsecs as officially determined by professional astronomers. So, for example, if we moved our Sun out to this pre-determined distance it would shine only at an apparent magnitude of +4.8. A red giant star like Betelgeuse in the constellation of Orion would have an absolute magnitude of −7.2, 60,000 times brighter than our Sun the same distance away, and more or less as bright as our Moon.

The Colour of Stars

When you look at the stars you will notice that they are not the same colour. They range from bluish-white to red.

The colour is largely determined by the star's surface temperature. Hot blue stars have temperatures of around 40,000 K. Cool red stars have temperatures around 4,000 K.

Astronomers classify the stars according to their chemical features which are determined through spectroscopic methods. The classification is as follows. O B A F G K M (W N S).

O type stars are denoted as the hottest and S the coolest. A useful mnemonic to remember this sequence is. 'Oh Be A Fine Girl Kiss Me right Now Smack!' even astronomers have a sense of humour!

Each spectral class is split into further sub-divisions 0 to 9. Our Sun is a yellow star of spectral type G2.

Star colours are enhanced through telescopes and binoculars. Double stars systems in these instruments often reveal more than one colour for the different components of the system.

Seeing Conditions

Seeing conditions of the sky vary from night to night. A clear sky with stars that are visible does not necessary mean that the sky is good enough to resolve other celestial objects to great detail.

A star's twinkling is not due to its intrinsic properties but is caused by the Earth's atmosphere. Air currents in the Earth's atmosphere, which moves between the observer and the star, distorts the path of the light coming from the star. Hence, the star will appear to momentarily brighten or dim or jump around in an irregular manner. So, the best time to observe the sky is when the air is steady and the air currents are less turbulent in the early hours of the morning.

The 'seeing' is an indicator of the steadiness of the air and depends upon a number of factors.

- The flow of air currents many kilometres above the observer.
- Localized phenomenon, such as the air flow inside observatories or telescope tubes.

Obviously the first factor cannot be directly controlled by the observer. The second factor however, can be controlled to an extent. Astronomers go to great lengths to ensure that they get the best out of the seeing conditions on any particular clear night.

During the hot summer months the ground soaks up heat during the day and then cools down at night, by exchanging heat between the air currents near to the ground which then subsequently rise upwards. This can lasts several hours, well into the night. Astronomers observing during this period find that the seeing is likely to be poor and very turbulent. Objects look like they are 'boiling' and there are rippling effects. The rate of improvement of the seeing during the course of the night depends upon the surroundings where the observing is taking place. For example, observing on stone surfaces or concrete patios tends to provide poorer images than observing on grass surfaces, because the grass surface is more efficient than concrete at dumping heat into the surrounding air. The best time to observe is in the wee small hours when the ground has cooled down sufficiently, that the air flow is much steadier.

Astronomy as a Hobby

Table 2.2. The Antoniadi Scale

Scale Factor	Description
I	Very good seeing, no air rippling effects.
II	Moments of good seeing, some air rippling seen.
III	Fair seeing. Air currents more turbulent.
IV	Poor seeing with constant air rippling effects.
V	Very poor seeing. Object barely made out and difficult to resolve.

An indication of good seeing is where haze near the ground is present. This means that the air currents are stable and under these conditions good views of objects, such as Planets and stars can been seen.

Seeing conditions are normally recorded in an astronomers' nightly log. They often use the Antoniadi Scale (Table 2.2).

Another condition of the sky is called transparency. The transparency of the sky is an assessment of its clearness. Good transparency implies that the stars are at their most brilliant, and that faint objects such as galaxies can be seen.

Transparency also depends on the altitude. A star at the zenith (directly overhead) can be as much as a few magnitudes brighter than a similar star near the horizon. The haze near the horizon contributes to this light dimming. It is particularly noticeable in light polluted areas where the haze is so thick that it blots out almost all but the brightest stars.

Weather conditions as well contribute to the transparency of the sky. A sky condition that is sometimes observed is when there has been a recent downpour. The clouds disperse as the weather front moves across and the sky then clears. The sky colour of low altitude region close to the horizon becomes a fantastic blue colour, gradating to deep blue as you go from the horizon to the zenith. It is not the usual hazy light-polluted orange that is normally observed. The downpour has washed most of the dust particulates out of the atmosphere, leaving a very transparent sky. However, this does not normally mean that the seeing conditions are just as good as the weather conditions associated with turbulent air currents.

For recording observations (next section) astronomers tend to use a simple scale 1 to 10 to record the transparency of the sky. Where 1 is very poor conditions e.g. high thin cloud, 5 is moderate and 10 is perfect crystal clear conditions. Don't worry if you don't get the sky conditions measurements correct as it comes with experience.

Seeing conditions and transparency of the sky are constantly changing so the more time you go out to observe, the better the chances are that you will get an observing session with perfect seeing conditions. When that happens the sky will open up to you.

Recording Your Observations

Taking notes of nightly observations is not for everyone, but it does provide a record of what you observed during an observing session. The detail of information you record is entirely up to you. Some astronomers meticulously produce notes in great detail whilst others just plain observe without recording anything at all. Personally, I only record notes when I deem it necessary, such as noting an interesting feature on a Planet. I prefer to produce more detailed notes during major astronomical events such as Lunar Eclipses, Comets, Planetary Conjunctions or Meteor Showers.

If you decide to produce a record of your observing, then I would recommend that you produce short-hand style notes when you're at the telescope. You can then write them up later in greater detail whilst it is still fresh in your mind. Each observation record can be tailor-made to suit the type of observing you carried out at the time, such as simply observing objects or more advanced CCD imaging.

An example of an Observing record template is in Figure 2.5 which shows basic details such as the date, time and location as well as sky and seeing conditions. The template has been tailor-made for users of LXD telescopes, where details relating to the mechanics of the telescope, such as alignment or motor problems can be recorded. These notes may prove useful for solving set up problems during later observing sessions.

You may wish to dedicate an entire observing session to a single object such as a Planet. In this case you can use observing record templates which have circular discs on them for you to sketch details on. These templates are freely available from the British Astronomical Association.

You can produce your own observing template using any word processing package. Alternatively, there are various astronomy computer programs and on-line facilities available on the internet for recording amateur astronomical observations. Their websites are listed in Appendix D.

Getting More Information

Astronomy resources are vast and cater for all astronomers; from the beginner to the very advanced. You may already have in your collection numerous astronomical materials that you refer to on a regular basis. However, if that is not the case then it is important that you find the right book which covers all the basics of observing and relevant astronomical information, along with decent sky maps.

Even if you use a Goto telescope that stores thousands of celestial objects in its handset, a good book is essential in order to help you find your way around the night sky.

Current up-to-date information such as forthcoming eclipses, occultations, planetary conjunctions, telescope reviews and spaceflight information can be found in astronomy magazines such as *Astronomy, Sky and Telescope, Astronomy Now* and *The Sky at Night Magazine*. These are available on many newsstands or from subscription direct from the publishers. The magazines are normally published monthly and directed to a wide audience whose astronomical experience ranges from the beginner to the advance amateur astronomer.

Other resources such as the internet are vast and you can spend many hours (and days) finding sites that contain relevant material. All aspects of astronomy are covered on the internet, including user groups and bulletin boards where astronomers talk and swap useful tips. A list of websites is in Appendix D.

Astronomy as a Hobby

Session Number:		Date:
Location:		

Set-up checklist:
Polar Align ❑   Goto Setup: Easy ❑   * ❑   ** ❑   *** ❑

Comment:

Seeing Conditions: Sky   Conditions:

Observations

Object	Time	Eyepiece	Comments

Planetary Details

Figure 2.5. An observing record.

Join an astronomical society. There should be a local one near you. Many are starting up as the hobby becomes ever more popular. Most societies cater for all types of experiences. You will find talking to someone and sharing your interest, attending guest lectures and going to star parties an enjoyable experience, not just from an astronomical perspective, but socially too. You can also join formal affiliations such as the American Astronomical Society or the British Astronomical Association.

As you can see there are many resources at your disposal to help you learn the night sky. Knowing that there are others out there who share your interest makes the hobby that much more worthwhile. You are not alone!

Conclusion

Astronomy has recently been enhanced through the use of technology, making it increasingly easier for observers to find their way around the sky. Sometimes though, you just simply cannot beat going outside on a clear night with nothing more than your mark one eyeball, gaze up into the night sky and see what it reveals.

Astronomy opens our minds up to endless possibilities and makes us ask the big questions such as 'Why are we here?' and 'What is our place in the grand scheme of things?'

So go out at night, look up at the sky, and imagine yourself floating among the stars. After all, we are made of starstuff!

CHAPTER THREE

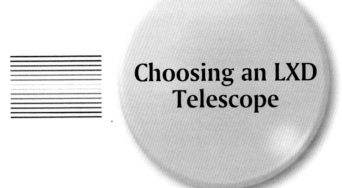

Choosing an LXD Telescope

When I was about 11 years old my interest in Astronomy had grown so much that I was encouraged by my parents to pursue it further by buying for me books and videos on the subject. One day, for my birthday they took me to a photographic shop in London to buy a telescope. When we walked into the shop, what I saw displayed in the corner made my heart pound and my senses whirl. There were four or five telescopes, on big tripods, looking like they had just landed from outer space! I was thrilled to bits that I was about to own my first ever telescope.

That night my dad and I set up the 4.5 inch reflector outside. I instantly fell in love with the night sky. Whenever it was clear or partly cloudy, I would be out there looking at the Moon and Stars or trying to find a bright Planet. The telescope wasn't the best instrument. Its tripod was wobbly and the eyepieces were poorly constructed, but I loved using it nevertheless. Alas, I don't own that telescope anymore, but since then I have owned many astronomical instruments. Some were better than others, but I still had many hours of enjoyment out of them.

When my parents bought me that telescope all those years ago, they never considered all the little problems or weaknesses that came with it. They were not astronomers, nor had they any experience of buying an astronomical instrument. It just looked impressive in the shop and that's why they decided to buy it. Those new to astronomy tend to purchase telescopes this way without seeking proper advice first.

The LXD series of telescopes is based on several different types of telescope designs. So which telescope should you choose? For a beginner it can be a daunting task choosing the right telescope. Even for amateur astronomers choosing their next telescope, careful thought is required.

This chapter provides a general description of the different types of designs used by LXD series of telescopes. The aim is to provide guidance on which type of telescope you should choose, in particular, from the LXD series. A summary description of the LXD telescope series is at the end of the chapter.

Tools of the Trade – Introduction to Telescopes

Telescope designs have come a long way since its invention in the sixteenth century. Over the centuries designs of telescopes have evolved significantly. Nowadays there are many types of instruments on the market, ranging from basic simple constructions to fully-automated computerized models. Fundamentally, they all use similar principles of refraction with lenses and reflection with mirrors.

Meade has used four designs of telescope for their LXD series.

- Refractor
- Newtonian Reflector
- Schmidt – Newtonian
- Schmidt – Cassegrain

Other designs such as Maksutov-Cassegrains are also commonly used for other telescope models such as the Meade ETX however, there is no 'Mak-Cass' model for the LXD series so its details are omitted from the book.

Astronomers often describe telescope specifications using terms such as focal length or focal ratio. The focal length is the distance of the primary objective or mirror to the focal point. The focal ratio is the focal length divided by the diameter size of the primary objective or mirror. These two terms are used throughout this chapter.

The Refractor

When the telescope was invented in the sixteenth century, the original design was a refractor. It used a simple front lens as its primary objective. Refractor designs have evolved since then, in particular objective lenses are now made of superior materials which greatly enhance the quality of the image. Modern refractor objectives comprise of at least two components. Figure 3.1 shows a schematic diagram of the LXD AR Refractor.

Several years ago the largest refractor that was popular and affordable amongst amateur astronomers had a primary objective lens of only four inches. Larger apertures were available, but they were more expensive, difficult to manufacture en mass, and usually had long focal lengths, making the tubes very long and bulky to use. However, the technology for manufacturing objective lenses has improved such that it is now possible to construct large aperture refractors with relatively short focal lengths. So, apertures larger than four inches are still quite portable.

The primary objective of a refractor defines the quality of the instrument. Single glass lens refractors tend to suffer from an effect called chromatic aberration. This is

Choosing an LXD Telescope

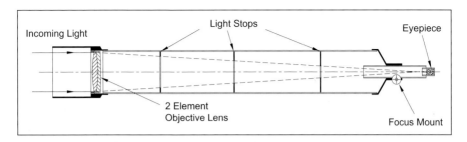

Figure 3.1. Schematic of the LXD AR refractor. Image courtesy of Alan Marriott.

when bright objects such as the Moon and Planets display false residual color effects around the edge of their discs.

There are a number of methods for reducing chromatic aberration effects in a refractor. One method is through combining lenses of different shapes and materials (achromatic doublets). Special surface coatings are often used. Most standard lenses are made of crown and flint glass elements placed together.

Filters are available which reduce the intensity of the false color, although they do not block it out completely. One of the lenses can be made of a different material such as Fluorite.

Refractors with primary objectives containing Fluorite are known as apochromatic refractors (APOs for abbreviation). APOs provide relative color free images of all bright celestial objects. Fluorite however does not come cheap and so the price to pay for quality of image is somewhat higher.

Nowadays achromatic refractors have been manufactured with very short focal lengths, some almost half the usual size. These are very portable instruments, but they suffer from chromatic aberration. In fact, the shorter the focal length the more severe is the effect. Just very recently, short focal length telescopes have been re-engineered and enhanced to include fluorite in their primary objectives, making them apochromatic. The reason for a short APO or semi-APO is that you get a high quality instrument which is extremely portable. However, the price is markedly higher compared to their achromatic counterparts.

When an object is viewed through a refractor it appears upside-down. This is because the primary objective lens at the front of the telescope inverts the image. This can sometimes be confusing as it seems counter-intuitive when moving the telescope in different positions i.e. when the telescope is moved in one direction, the object seen in the eyepiece moves in the completely opposite direction in the field of view. There are accessories available that invert the image the 'right' way up; however, they are more appropriately used for viewing terrestrial objects rather than astronomical objects. To help the observer become familiar with inverted images of astronomical objects many astronomy books tend to show maps of the Moon and images of Planets with south at the top just as if they were viewed through a refractor.

Refractors tend to be mainly used for solar system observing rather than deep sky observing such as Galaxies or Nebulae. Their long focal lengths provide high magnifications and small field of views. Hence they can resolve fine details of the Moon and Planets without significant loss of contrast and sharpness.

A disadvantage of a refractor is that the primary objective lens tends to 'dew' up in cold weather thereby reducing the quality of the image. Dew is when tiny water droplets form on the surface of the lens. This happens when the lens cools down faster than the surrounding air. Dew shields are used to increase the time before the objective dews up, but electronic heaters are usually more effective at keeping the objective dew free. We will discuss methods for handling dew later in this book.

The Newtonian Reflector

The reflecting telescope became the rival instrument to the refractor when it was first constructed for practical use by Sir Isaac Newton back in the seventeenth century.

A Newtonian telescope consists of a concave primary mirror and a small reflective flat mirror located at a fixed distance from the primary. Light reflects off the primary mirror and converges to a focal point a distance away determined by the focal length of the mirror. The *secondary* mirror or *flat* is positioned at a 45° angle, which deflects the light from the primary mirror to the side of the tube into a focuser (Figure 3.2).

Large aperture reflectors make them essentially a big *light bucket*. The greater the diameter of the mirror the more light gathering power it has. This makes it an ideal instrument for observing deep sky objects such as Galaxies and Nebulae.

Most reflectors have typical focal ratios between f/4 and f/8. 'Fast' optics with focal ratios of f/4 or f/5 tends to keep the length of the telescope tube as short as possible, thereby making them highly portable. These *fast* optics are suitable for observing deep sky objects as they tend to provide wider fields of view than their counterparts, which have slower optics of focal lengths f/7 or higher. The higher focal ratios tend to be more ideal for planetary observing as their field of view is smaller, and have the ability to provide higher magnification. However, telescope tubes for these high focal ratios tend to be longer in length, and are not as portable as the shorter tubes. Moderately sized mirrors diameters between six and ten inches typically have focal ratios of around f/6, providing the best of both worlds for deep sky and planetary observing.

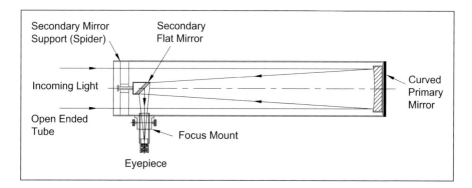

Figure 3.2. Schematic of the LXD N-6 Newtonian reflector. Image courtesy of Alan Marriott.

A disadvantage of a reflector is that it tends to suffer from air currents within the tube because of its open front-ended design. This effect becomes more evident as the aperture size increases. When the telescope is taken out of a warm house and used for the first time at night, images of objects behave as if they are 'boiling', and the quality is severely degraded. As the tube cools down after a period of time, the image starts to improve when the air currents disappear inside the tube, due to the internal temperature equalizing with the external environment. The time that the cooling process takes depends upon the construction of the tube itself. Aluminum metal tubes tend to fair better and cool down to optimum temperatures quicker than rolled cardboard tubes. This waiting time could be a hindrance, especially if the observer wants to try and get in a quick observing session before the weather changes for the worse.

Nowadays, telescopes are designed to reduce the time it takes for the telescope tube to cool down. Designs include, sealing off the tube with a front transparent plate (see Schmidt telescopes in next section), placing baffles at regular intervals along the inside of the tube and, using fans to cool the primary mirror. These solutions are very effective at reducing the air currents within the telescope tube, as well as improving the quality of the image. A well-engineered reflector tube assembly can compete with some of the best refractors on the market in image contrast and sharpness.

Schmidt-type Telescopes

The Schmidt system belongs to a group of telescopes called Catadioptrics. These telescopes use a thin and a near-flat front lens called a corrector plate. A secondary mirror used is not flat like the ordinary Newtonian design. It is convex designed to correct aberrations produced by the telescope's spherical primary mirror. The secondary mirror is attached to the centre of the corrector plate. In fact, the mirrors and the front correcting plate are specially designed in order to achieve the best quality images from the Schmidt configuration. The front plate also protects the mirrors inside the tube from corrosion from constant exposure to weather elements, every time the telescope is used.

There are currently two common types of telescopes on the market which employ the Schmidt system:

- Schmidt-Newtonian
- Schmidt-Cassegrain

Schmidt-Newtonian

An enhanced version of the classic Newtonian design is the Schmidt-Newtonian telescope (commonly abbreviated to SNT or Schmidt-Newt). The SNT is a relatively new design that has recently become popular. Many manufacturers have tried producing them in the past, but constructing and configuring the optics proved to be difficult, and the resultant quality of the images were poor. Nowadays, manufacturers seemed to have resolved the issues for constructing SNTs en masse.

The SNT employs a Catadioptric design which uses a specially shaped front lens as well as mirrors arranged in a standard Newtonian configuration (Figure 3.3). It is essentially a cross between a Reflector and a Refractor. SNTs are superior optically to

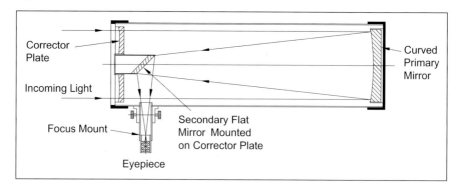

Figure 3.3. The LXD SNT design. Image courtesy of Alan Marriott.

standard Newtonians. The mirror does not need to be a perfect shape (rather near perfect), since the front lens corrects any aberrations produced by the primary mirror.

Using a front corrector plate enables typical focal lengths shorter than standard Newtonians. Hence, these 'fast optics' telescopes imply they are very much suited for deep sky observing and in particular, astrophotography and CCD imaging.

A closed tube configuration means that SNTs do not suffer from poor image quality due to turbulent air currents inside the tube. However, they take longer to cool down than their standard Newtonian counterparts. Moreover, the tube assemblies of the SNTs are heavier than standard Newtonians, since the front corrector plate has to be of a sufficient thickness. This is to provide satisfactory correction for aberrations of the main mirror.

Thanks to improvements in technology to enhance the quality of the optics and mass production, SNTs are becoming the affordable compromise for amateur astronomers, although they probably will never be as popular as Schmidt-Cassegrain telescopes.

Schmidt-Cassegrain

The Schmidt system is combined with a Cassegrain design to produce the Schmidt-Cassegrain telescope (commonly abbreviated to SCT or Schmidt-Cass). The SCT has been around for many years and primarily took off in the early 1980's. It soon became very popular amongst amateur astronomers. SCTs have had much more commercial success than SNTs as they are relatively easier to produce.

The classic Cassegrain telescope is somewhat different to the Newtonian design. Firstly, the primary mirror is not parabolic, but spherical in shape. These spherical mirrors are easier to manufacture than the parabolic design. However, spherical mirrors suffer from spherical aberration, so telescopes based upon these mirrors incorporate specially shaped convex secondary mirrors in order to reduce the aberration.

Secondly, the primary mirror has a hole in its centre to allow light to pass through that is reflected from the secondary mirror. The focal area is positioned directly behind the hole of the primary mirror, which is where the light from an object converges and is focused.

Choosing an LXD Telescope

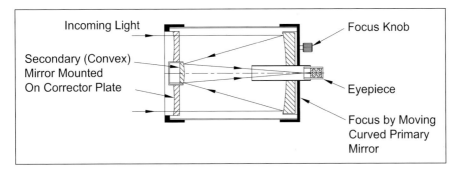

Figure 3.4. The LXD SCT design. Image courtesy of Alan Marriott.

An interesting design feature, as shown in Figure 3.4, is to focus on an object, the primary mirror moves along the tube. This is different to the way a standard focus mount is used (discussed later in this chapter) nonetheless the final result is the same, a smooth accurate focus. Older designs of SCTs used to suffer from 'image shift', where the mirror was not perfectly maintained concentrically within the optical path. When an image was being focused, it would appear to marginally shift to the left or right, depending upon where the mirror was with respect to the center of the optical path. I have not heard of anyone having this problem anymore with recent SCT designs. The convenience of having a focus at the back of the telescope makes it far easier to view objects in most positions. This is in comparison with a reflector, whose focuser is situated near the top side of the tube. Hence, for large apertures with very long focal lengths a ladder is possibly required to access the focus.

An advantage of an SCT is that multiple mirror configuration makes the focal length quite long for a relatively short telescope tube design. The light has to travel several times up and down the tube, reflecting off two mirrors, before reaching the focus area at the back of the telescope. Focal ratios are typically f/10. Hence, these telescopes are excellent for planetary work as well as deep sky observing. This is what makes these types of telescopes so appealing to astronomers, although they tend to be more expensive than other telescopes. Optical adapters are available which, when attached to the back of the tube at the focuser, reduces the focal length by changing the focal ratio from f/10 to f/6.3 or even f/3.3. This makes the SCT a versatile instrument, providing excellent quality images of planetary and deep sky objects, as well as being an ideal instrument for astrophotography.

A disadvantage of an SCT is that the front plate tends to suffer from dewing like all other Catadioptric designs.

Telescope Tube Essentials

I will be discussing more about accessories for telescopes in Chapter 12, but for now, I will talk briefly about focusers and finderscopes, the two most essential components of a telescope tube set up.

Focusers

The focus mount is the 'business end' of the telescope tube assembly. It is an intrinsic part of the telescope, where eyepieces are used in order to view objects at various magnifications. A typical focuser is split into two elements; the drawtube and the focusing rack.

Most telescopes have focusers that employ the rack and pinion method for focusing. The action of twisting the focusing knobs moves the drawtube in and out of the telescope via a toothed gear, thereby providing a focus of the object in the field of view. Other focusers are available that provide superior features than the rack and pinion, such as ultra-smooth focusing and zero sideways shift capability (Crayford Mounts).

Focusers come in three standard drawtube diameters:

- 0.965 inch (24.5 mm)
- 1¼ inch (31.7 mm)
- 2 inch (50.8 mm)

The 0.965 inch focuser (commonly termed as 'one inch') is usually supplied with cheaper branded models. Adaptors are available which allow 1¼ inch eyepieces to be used with these cheaper brands. However, they do not really improve performance of the images, since the quality of optics are likely to be poor in the first place.

The 1¼ inch focus mount has been included with almost every telescope that has been manufactured over the past several decades. Hence, it is also the most common size for which most eyepiece barrels are based upon.

The 2 inch focuser is a relatively recent addition to the focuser range, especially for entry-level telescopes purchased by those new to astronomy. They are now available on many models, and normally come with adaptors which allow the use of 1¼ inch standard size eyepieces. Typical fields of view with 2 inch eyepieces are generally wider than their 1¼ inch counterparts and the low power eyepieces almost provide binocular-type fields of view.

The mark of a good focuser is the preciseness and smoothness of operation. When you twist the focusing knobs, you should sense the drawtube moving smoothly in and out of the focus barrel. There should be no lateral movement sideways either, evident by observing the image shifting sideways in the eyepiece when it is being focused.

There are many different designs of focusers on the market. The best focusers typically are low profile in design so that they don't protrude out of the telescope tube assembly too far. If the focuser and drawtube is very long, attaching accessories such as long focal length eyepieces, Barlow lens or CCD cameras will likely overbalance the telescope at the focusing end. Hence, make it difficult for the telescope to operate smoothly. Low profile focusers are also best for astrophotography and CCD work, as they allow for prime focus imaging. This generally requires access closer to the tube.

In most cases where telescopes are sold as complete systems, you cannot choose the focuser type, so you have to compromise and be satisfied with what you get. On the whole though, most focusers are of sufficient quality for observing, and do not often require replacing unless they are in need of repair or the observer requires something more specialized.

Choosing an LXD Telescope

Finderscopes

It is actually not as easy as you might think to find an object through a telescope without some kind of lower power wide-field optical aid. This is due to fact that telescopes have relatively small field of view, so it is difficult to line up the telescope tube at an object and see it in the eyepiece.

A finderscope fits to the side of the telescope tube. It is designed to provide a low-power wide field of view in order to aid the observer to find objects in the sky with the main telescope. This is done by aligning the finderscope in a parallel direction with the main telescope, such that what you see in the finderscopes' field of view is also seen in the field of view of the eyepiece.

The most popular finderscope model is the refractor type, which consists of an objective about 30 to 50 mm in diameter and a small eyepiece with a magnification of $6\times$ to $10\times$. Looking through the finderscopes' eyepiece, you will see a set of cross-hairs. These are thin-wires set such they traverse the diameter of the eyepiece and intersect at the center of the field of view.

Alternative cross-hair arrangements employ double-stranded wire or concentric circles, directly etched onto the back of the eyepiece glass at the centre of the field of view. The patterns are usually of the single strand variety as they allow the user to align an object within a small square or circle area, rather than it being obscure behind thick strands that intersect. There are various different types of finderscopes available. These are discussed in Chapter 12.

Telescope Mounts

Choosing the right mount is just as important as choosing the right telescope. Nowadays, magazine advertisements appear to emphasise the features of telescope mounts just as much, if not more so, than the tube assemblies.

Telescopes are mounted in various ways using a variety of designs of mounts. There are mounts which allow the telescope to track the stars, while others simply allow the user to point to an object and manually track it themselves. Mounts that incorporate computerised technology provide setup features with little or no user intervention. This can be advantageous, as it speeds up the time to set up the telescope at the start of an observing session.

Most beginners when choosing a telescope normally consider the telescope tube only. They tend to discount the importance of the mount thinking that it's just there to provide support for the telescope, and assume that it's good enough to find objects in the night sky. If the mount is not up to standard, the user will end up disappointed and frustrated at the poor performance of the instrument and may decide not to use it anymore.

It is very important that you choose the right type of mount when you buy a telescope. It is no good having quality optics if the mount lets you down. Time and time again those new to astronomy purchase instruments with good quality, but are let down by a wobbly, unmanageable telescope mount. A good sturdy mount aids viewing by producing a steady image through the telescope. Mounts that vibrate at the slightest movement tend to provide poor images. This is because even though the human eye has the ability of resolving detail on moving vibrating images to a degree, it has the ability to resolve much higher detail if the image is still or steady.

There are tests for checking the quality of a telescope mount. These tests will provide a good indication of the quality of the mount and whether it can manage the tube assembly during observing sessions:

Vibration Stability Test. A common method for testing the stability of a mount is the two second test. Set the telescope up with all components assembled and momentarily tap the telescope tube assembly whilst looking through an eyepiece at an object. You should be able to see the image vibrate and wobble. After a few seconds the vibration should die away, leaving a stable image again, which is a sign of a good mount. Poor design mounts not only will vibrate at the slightest movement, but also flexes at the hilt of the stand, if heavy tube assemblies are mounted on them.

Movement Test. A good sign of a decent engineered telescope mount is whether the movement of the telescope in any orientation is smooth and responds to very small shifts in motion. Rough, jerky movements are a sign of badly engineered worm-wheels and gears, or the gears need to be worn in with use. Either way it is not a good sign of quality and will result in the user being frustrated, especially when they are trying to keep an object in the centre of the field of view at high magnifications.

Mount Types

A telescope mount is considered to consist of two main components; the mount 'head' (commonly termed as the 'mount'), and the telescope stand. The mount head controls the movement and orientation of the tube assembly. The stand provides the support for which both the mount head and tube assembly are reliant upon for stability.

Mounts come in a variety of models, all of which are intended to help the observer in locating objects in the night sky and track them. This can be done manually through turning slow motion control gears by hand, or automatically through the use of electric motors.

There are fundamentally two types of mount:

- Altazimuth
- Equatorial

Altazimuth Mounts

The Altazimuth mount is the simplest design of all. Two axes provide horizontal (azimuth) and vertical (altitude) motions. There are no special complicated alignment or setup procedures before you can use it. All you need to do is make sure that the tripod and mount are level with the ground. Objects are found by moving along the horizontal axis and then along the vertical axis a specific distance. Objects can be tracked for short periods by making small incremental steps horizontally and then vertically. The mount is not suitable for long period tracking, as the field of view containing the object being tracked will tend to suffer from a rotational effect. The effect is because the Earth's axis is tilted to one side. Hence objects in the sky do not traverse the sky in horizontal or vertical motion, but at an angle depending upon

Choosing an LXD Telescope

where you are. So, you will see the object at different orientations in the telescope's field of view as it moves across the sky, although it is not very noticeable. An equatorial mount compensates for this effect.

Small refractors commonly use the Altazimuth design and are ideal for beginners new to observational astronomy. Large amateur astronomical telescopes such as Dobsonians make use of the Altazimuth design, as well as smaller fork mounted telescopes such as the Meade LX200 or Meade ETX. Some of the largest telescopes in the world use an Altazimuth mount.

Equatorial Mounts

The equatorial mount is designed to track objects in the sky for long periods of time. The equatorial mount is basically an Altazimuth mount tilted towards an angle such, that the altitude axis is aligned in parallel with the North Celestial Pole, and the azimuth is in parallel with the Celestial Equator (hence the name of the mount). In this orientation the azimuth axis becomes the RA axis, and the altitude axis becomes the Dec axis (see Chapter 2 for explanation of RA and Dec).

There are a number of different mount designs which are used by amateur astronomers, but the most common one used nowadays is the German Equatorial Mount (GEM) (Figure 3.5). GEMs consist of two axes at right angles to each other, which constitute the RA and Dec axes. A set of counter-weights is placed opposite the tube assembly at the other end of the Dec shaft in order to provide a counter-balance. Fixed to each of the axes are round discs with numerical gradations or notches etched on them called Setting Circles. These discs use the RA and Dec coordinate system and are used to locate objects in the night sky. Once the setting circles are calibrated, an object in the sky can be located by simply 'dialing' up its respective coordinates. Nowadays,

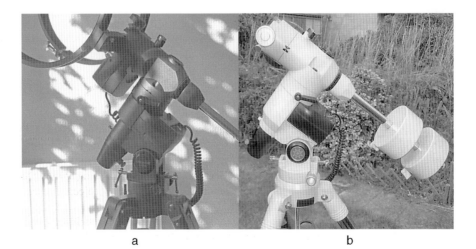

a   b

Figure 3.5. The LXD GEM mount. (a) LXD55, (b) LDX75.

digital setting circles are used in electronic handsets, which mean that the setting circle discs are not used often anymore.

If an electronic motor is fitted to the RA axis of the GEM it will drive the axis at Sidereal rate in the direction which counteracts the motion of the stars. In other words, the mount will be able to track the stars for lengthy periods. Hence, it is easier to carry out detailed observations of celestial objects without the need to make constant RA adjustments.

For those who are used to using Altazimuth mounts, it takes practice and some patience to handle an equatorially mounted telescope. Moving the RA and Dec axes into particular positions can seem counter-intuitive at times and sometimes the mount doesn't seem to behave as it should. However, it doesn't take long to master the axial motions.

The more serious amateur astronomer will wish to use an equatorially mounted telescope in order to carry out long time exposure astrophotography or deep sky CCD imaging. These tasks require precise polar alignment and this can increase the amount of time it can take to the set the telescope up.

Technology has assisted us with improving the design of equatorial mounts. They are now made of superior materials that can sustain wear and tear, as well as providing improved tracking than their predecessors 15 to 20 years ago. GEMs are mass manufactured and exported primarily from China. Despite having different model names such as EQ5, Celestron CG-5, Vixen GP and of course the Meade LXD, they are all based on the same fundamental GEM design, with slight variations to suit each manufacturers' model.

The new designs include latitude adjusters as well as built-in polar alignment scopes, which in my opinion, are probably the most two significant enhancements that have been introduced to GEMs to date. Latitude adjusters are highly desirable amongst users of the new mounts. This is because most equatorial telescope mounts that were on the market several years ago, had a single large bolt through the centre of the mount. Undoing this bolt was the only way to change the latitude of the mount. This proved extremely difficult to adjust, especially if the telescope was taken to different latitudes. The bolt had to be sufficiently tight in order to prevent the telescope's weight from accidentally changing the latitude when it was pointed in different directions, or when counterweights were added to it.

Further technological enhancements means that the Equatorial mount can be fitted with a Goto system, just like the Altazimuth mount. This makes equatorial Goto systems very appealing to amateur astronomers who do not wish to compromise precise tracking with Goto capability.

Telescope Stands and Tripods

A telescope stand's function is to provide support and stability for both the mount and optical tube assembly. The two most common types are the tripod stand and the pillar or pedestal stand.

Portability plays an important factor for which type of stand is used. For example, if you are regularly taking the telescope outdoors or to different locations to observe, you will prefer to have a stand that is relatively lightweight and quick to set-up, as well

Choosing an LXD Telescope

as being reasonably sturdy enough to manage the combined weight of the telescope and its mount. A tripod would be the preferred stand in this case. However, if you are using a telescope at a fixed location (such as an observatory), or the combination of the telescope and mount is quite heavy, then a pillar stand is more suitable.

As an owner of many different types of telescope stands over the years, I would say that a robust pillar stand provides better stability than a tripod stand. This is evident when observing an object in the eyepiece of an instrument mounted on a pillar stand. The image in the field of view is not as susceptible to vibration, compared to using a tripod stand. Thus, the eye can track the image better, revealing more details of the object. I would point out, that tripod stands nowadays are constructed with materials that are very rigid. Some of the larger tripod stands can be almost as good as a pillar stand for handling heavy instruments.

Telescope stands can vary in cost, and you will have to weigh the pros and cons within the budget you can afford. Ultimately, you have to choose a stand you think will be suitable for the telescope you will eventually use.

Tripod Stands

The majority of telescopes with GEMs have tripod stands. A tripod stand typically consists of three legs separated at equal distances from each other. The legs are usually constructed with long pairs of rectangular struts that are usually made out of wood or aluminum. The aluminum struts are normally hollow in order to reduce the overall weight of the tripod. Some designs have legs which are tubular and made out of stainless steel. These tripods provide vastly superior stability than their wooden or aluminum counterparts. All three legs are connected to an adaptor plate. The telescope mount is fixed via a single threaded bolt and passes through the plate into the underside of the mount.

The legs of a tripod are normally adjustable in height so that they cater for telescope tubes of varying lengths. Refractors and Schmidt-Cassegrain telescopes should have the tripod legs set to their maximum height. This is to achieve a comfortable position for the observer looking through the eyepiece at the bottom end of the telescope. A reflector having the focus mount located near the top of the telescope tube, should have the tripod legs set at a lower height, to provide more comfortable viewing.

Setting up a tripod is more or less straight-forward. Minor adjustments to the height of the legs are necessary to make sure the telescope and mount assembly is level with the ground. All three legs must be adjusted until leveling is achieved. Leveling the tripod is easier to achieve on flat surfaces such as concrete paths and patios, than uneven ground such as gardens and inclines (see Chapter 4 on how to set up a tripod). Some tripods come with notches or numbers imprinted onto the legs at regularly spaced intervals in order to speed up the process, as it can be a time-consuming task trying to obtain a perfect level. For basic Goto operations only a rough leveling is necessary, so you shouldn't waste too much time on the task if you are going to be moving the telescope in between observing sessions.

Many tripod stands have an accessory tray, like the LXD55 and LXD75 tripods where eyepieces and other accessories are placed on ready to be used during an observing session. These trays are normally fixed to all three tripod legs and enhance the overall stability of the stand.

Figure 3.6.
LXD75 tripod mount.

There is a drawback with tripod designs that have aluminum or wooden struts used as legs. The top of the legs are normally attached to the plate using single wing bolts. Owners of these tripods have pointed that when the tripod stand is under stress from the weight of the telescope, the points where the struts are attached to the adaptor plate, tends to 'flex' and 'twist' slightly when the telescope tube is moved around in different positions. This could lead to inaccuracies in finding objects using the Goto facility. The solution is to have better designed tripods, which reinforce the strength of the legs by having central leg clamps or using legs of a tubular design made out of rigid materials such as stainless steel. This is the case with the LXD75 models (Figure 3.6). See LXD Tripod section later in this chapter.

Quality wooden constructed tripods tend to handle vibrations better than the hollow legs of an aluminum tripod. They are usually aesthetically pleasing, made of polished oak or other types of high quality wood. However, wooden tripods tend to weigh more and require more looking after than their metal equivalents. Aluminum tripods are less prone to wear and tear during transportation and are not affected by weather and temperature differences, unlike the wooden tripods.

Choosing an LXD Telescope

Figure 3.7. Typical pillar stand. Image courtesy of Alan Marriott.

Pillar and Pier Stands

Many telescopes between eight and fourteen inches in aperture size use pillar stands (Figure 3.7). These stands are more rigid than tripods and can handle much heavier telescopes.

A pillar stand consists of a rigid heavy-duty single vertical pillar made out of cast iron or aluminum. Three or four support 'feet' are attached to the lower end of the pillar. The feet can be detached from the pillar so that the stand can be dismantled and transported elsewhere if necessary.

A disadvantage of a pillar stand is that it can be cumbersome to set up if it is not permanently fixed at a location. It normally takes longer to assemble the feet sections, unlike a tripod which can fold down in one step. The stand is not usually transported in one single piece due to its size and sheer weight when fully assembled. However, this does not deter those who are eager to transport these stands to any location in order to provide the ultimate stability for their instruments.

A permanent type of pillar stand is the pier stand. A pier stand is primarily used in amateur observatories or locations where a permanent base is required. The pier is made of either a concrete or brick constructed post, or a pillar type stand bolted to the floor and has the facility to attach as telescope mount to the top of it. Construction is such that the mount can be detached from the pier for maintenance when necessary. A disadvantage with a permanent mount is that if you decide to move residence, then you will probably have to build it all over again.

Personal Factors to Consider when Choosing a Telescope

There are many factors to consider when thinking about purchasing a telescope:

- Purpose
- Quality and Cost
- Portability

Purpose

When people ask me what sort of telescope they should choose as a first instrument, I tend to respond with the reply, 'What will you use the telescope for?'

For a beginner this question may seem difficult to answer, especially if they have never carried out any type of detailed observations through a telescope before. Experience helps in these matters, and amateur astronomers tend to find it easier to choose their next telescope than beginners their first. However they still have to ask themselves that same question, 'What do I want to use the telescope for?'

When you look at the sky you see a whole variety of different objects and as a result there are several different fields of astronomical observations that you can undertake. You may decide that you want an all-rounder instrument to carry out any types of observation or alternatively you may go for an instrument that is best suited for a particular type of observational work.

Typically, astronomical observations fall into the following categories:

- Solar System
- Stars
- Deep Sky
- Imaging (Traditional and Digital)
- Specialized

Solar System Observing

The Sun, Moon and the Planets are fascinating objects to study. Telescopes with good quality optics and those that produce high magnifications are required to provide sharp images of planetary discs, with minimal loss of brightness or contrast.

Choosing an LXD Telescope

The LXD Refractor and SCT make ideal planetary, Solar and Lunar telescopes due to their ability to produce high magnification, narrow fields of view and excellent contrast.

Observing Stars

Stellar astronomy is associated with the study of the stars such as binary systems, determining colors of stars, and stars that vary in brightness.

These objects require high magnifications in order to split stars that are in close visual or apparent proximity to each other. However, most astronomers don't use their telescopes just to study the stars, but use them for other purposes as well. Hence, telescopes used for planetary work are best suited for this type of observing as they can provide the greatest resolving power such as the LXD Refractor and the LXD SCT.

Deep Sky Observing

Deep Sky astronomy deals with objects such as Galaxies and Nebulae as well as faint comets. Most deep sky objects do not appear as star-like entities, but have real apparent surface area projected in the sky. This means that these extended objects usually have low surface brightness. Astronomical books usually refer to their brightness as if the objects were concentrated into a single point. This can be deceiving, as the object in question is usually much fainter than suggested, so could easily be missed by the observer.

To view these faint elusive objects, as much light gathering power as possible is required. Where small aperture refractors struggle to find those faint elusive deep sky objects, large aperture telescopes have no problems in finding them at all. Telescopes such as the LXD SNTs with relatively large apertures and wide fields of view make ideal instruments for this type of observing. The LXD SCT is also widely used for observing deep sky objects due to its aperture size.

Image Capturing

Capturing images of objects in the night sky using conventional 35 mm Cameras or CCD electronic imaging requires the parent instrument to be able to track the stars in a precise manner. Moreover, instruments with optics that enhance the brightness of faint deep sky objects, termed 'fast optics' are necessary in order to make the exposure time as short as possible. These 'Fast optics' are usually associated with reflecting telescopes with short focal lengths (see Reflector section later in this chapter).

The LXD SNT is one such instrument with 'fast optics'. Incidentally, the optics of the LXD SCT can be made 'fast' if special adaptors are attached to it.

Specialized Observations

This type of observing covers specialized astronomical topics such as Spectroscopy or Photometry. The object that is studied is usually a star or planet. 'Fast' optics would be an advantage in order to gather as much light as possible to obtain bright spectra. However, these specialist disciplines in most cases require dedicated equipment or specific modifications to the instrument as well as the right amount of planning and patience in order to achieve the desired results.

Quality and Cost

Another major consideration when choosing a telescope is the price. It is important to know when equipment purchased is value for money. For a beginner the affordability of a telescope is probably the most important factor when buying such equipment, especially if it is their first telescope.

The 'quality' of telescopes has somewhat improved over the past 10 years and the old adage "You get for what you pay for!" is not necessarily the case anymore when considering telescopes in the mid-price range $600 to $2000 (£500 to £1300). Nowadays, a relatively 'cheap' telescope could still have good quality optics. Of course, if you still want a high quality engineered instrument with the best optics then you still need to pay out much more.

The number of features available on a telescope per cost is also something to consider. The features on a telescope determine how the telescope will be used. There are many telescopes on the market with features that make them most popular amongst astronomers to purchase, i.e. Goto mounts, built-in Global Positioning Systems, Enhanced Optical coatings etc. A compromise has to be made if a budget has to be adhered to.

You should also keep in mind that once a telescope has been purchased, the cost of additional accessories must also be taken into account. Hence, the overall amount could be higher than the original expected outlay. The LXD telescopes are what I would term as 'mid-entry' level telescopes on account of the affordable features available.

Portability

Telescopes come in various shapes, sizes and different weights. It is important to consider the location where the telescope will be used. For example if a telescope is going to be taken out of storage every night and assembled every time then the telescope's size and weight becomes very important. It's no good buying a top-of-the-range huge and heavy telescope if it takes hours (and a hernia!) to carry it outside to set it up.

What is needed is a telescope that is relatively lightweight without compromising stability and shaped such, that carrying it outside is manageable. The counterweights and the mount components are typically the heaviest to handle. The telescope optical tube is the bulky part and special care needs to be taken when handling it.

The LXD Telescope Series

Now I have discussed the types of instrumentation on offer let's talk briefly about the LXD telescopes.

Choosing an LXD Telescope

Table 3.1. Telescope Models of the Meade LXD Series

Model	Series Available	Aperture Size	Focal Length	Telescope Type
AR-5	LXD55/75	5 inch (127 mm)	f/9.3	Achromatic Refractor
AR-6	LXD55/75	6 inch (150 mm)	f/8	Achromatic Refractor
SN-6	LXD55/75	6 inch (150 mm)	f/5	Schmidt-Newtonian
SN-8	LXD55/75	8 inch (203 mm)	f/4	Schmidt-Newtonian
SN-10	LXD55/75	10 inch (254 mm)	f/4	Schmidt-Newtonian
SC-8	LXD55/75	8 inch (203 mm)	f/10	Schmidt-Cassegrain
N-6	LXD75 Only	6 inch (150 mm)	f/5	Newtonian Reflector

Meade no longer produces the LXD55 models and only generally available through outlets such as auction sites, or private owners as second hand. The LXD75 however is currently available at the time this book went to press. A list of models in the series is in Table 3.1.

AR-5 and AR-6 Refractors

Both the LXD55 and LXD75 AR Refractors are achromatic in design. The primary objective lens consists of two elements (doublet) of different glasses in order to reduce chromatic aberration, amongst other optical effects.

The AR-6 has a focal length of 48 Inches which is over a meter in length. This makes an impressive sight for anyone who first sets eyes on it and certainly has the 'Wow factor' about it (Figure 3.8).

The AR-5 and AR-6 are ideal models to own if you want a telescope that produces excellent image contrast and resolution. They have the aperture size to optically perform on-par with other telescope designs having similar or larger apertures. Table 3.2 provides specifications of the LXD AR Models.

N-6 Newtonian Reflector

The N-6 model is a classical Newtonian reflecting telescope (Figure 3.9). It was introduced by Meade for the first time to the LXD75 series, to serve as a budget level telescope supplied without the Autostar handset computerized controller. However, you can purchase the handset for the telescope as an optional extra. A simple electronic controller (EC) is supplied with the telescope, which performs basic motor control functions. Table 3.3 provides specifications of the LXD N-6 Model.

The primary mirror of the N-6 is constructed out of plate glass unlike the SNTs or SC-8 which uses Pyrex. For Newtonian telescopes with large diameter mirrors, plate glass tends to cause problems with the focal length of the telescope. This is because the focal point marginally changes whilst the mirror is cooling down to optimum ambient temperature as the glass contracts in lower temperatures. Pyrex however, has low thermal expansion properties and is therefore suitable for large diameter mirrors. Meade probably decided that a 6 inch plate glass mirror would not have a noticeable

Table 3.2. AR Refractor Specifications

Model	AR-5	AR-6
Aperture:	5 inches	6 inches
List Price:	$1349.00	$1649.00
Supplied Eyepiece (4000 Series):	SP26	SP26
Highest Useful Magnification:	229×	305×
Visual Limiting Magnitude:	13	13.4
Focal Length:	1143 mm	1219 mm
Focal Ratio:	f/9	f/8
Resolution (arc seconds):	0.91	0.76
Net Weight:	49 lbs	72 lbs

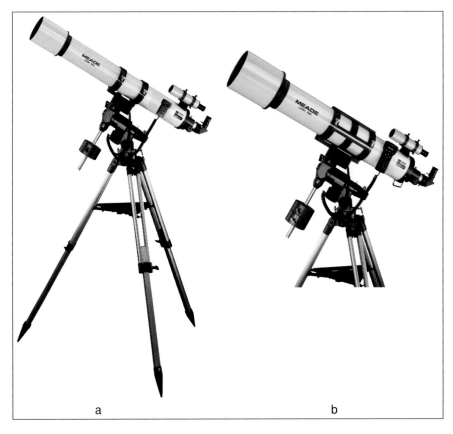

a   b

Figure 3.8. LXD55 AR refractor models. (a) AR-5EC. (b) AR-6AT. Images courtesy of Meade Instruments Corp.

Choosing an LXD Telescope

Table 3.3. Newtonian Specifications

Model	N-6
Aperture:	6 inches
List Price:	$899.00
Supplied Eyepiece (4000 Series):	SP26
Highest Useful Magnification:	254×
Visual Limiting Magnitude:	13.4
Focal Length:	762 mm
Focal Ratio:	f/5
Resolution (arc seconds):	0.76
Net Weight:	47 lbs

Figure 3.9. LXD75 N-6 Newtonian reflector model. Image courtesy of Meade Instruments Corp.

difference of the overall focal length with temperature differences, so they went with the plate glass mirror instead.

If you are new to observing then the N-6 Reflecting telescope is an ideal telescope if you want to learn the sky manually without the use of any automated Goto facility. An Autostar handset can be attached to the telescope, perhaps when you have gained enough experience and confidence to deal with a more complicated setup.

SN-6, SN-8, SN-10 Schmidt-Newtonian Telescopes

As discussed earlier in this chapter, the SNT has not been favoured by many manufacturers due to its complex design, especially with constructing the front corrector plate. Meade however has managed to produce an SNT system, whose optical performance has on the whole been classed as excellent by owners of the telescope (Figure 3.10).

The SN-6 is optically superior to the N-6 Reflector and yields better quality images as a result of the optical configuration. The relatively low weight and small aperture size of the SN-6 makes it ideal if you are looking for a portable instrument that has all the features of the LXD range. However, the size of the 6 inch aperture is not excessive and although the quality of images are very good, it does not have the light grasp to provide views of faint deep sky objects, compared to its bigger brothers the SN-8 and SN-10.

The SN-8 has over 50% more light grasp than the SN-6 and is best suited for viewing deep sky objects due to its 'fast' focal length f/4. Its size implies that it is portable although it is heavier than the SN-6.

The SN-10 is the largest of the SNT group. It literally is a 'light bucket', with over 50% more light grasp than the SN-8, and at f/4 it is perfect for observing and imaging very faint deep sky objects and Planets. Some owners of the telescope have stated that the LXD55 mount struggled with operating the SN-10, due to its sheer weight. However, the LXD75 mount fares better, due to the mount being re-engineered to handle the bigger models of the LXD series.

Meade has found a niche in the market for these types of telescopes, and combining them with features, such as UHTC optics and Goto, they have put them on many astronomers' wanted lists. Table 3.4 provides specifications of the LXD SNT Models.

SC-8 Schmidt-Cassegrain Telescope

Meade has been producing SCTs for a long time, so it was a logical decision that an SCT would be introduced into the LXD telescope series (Figure 3.11).

At f/10 the SC-8's focal length is quite long (2 m). However, the SC-8's folded optics configuration greatly reduces the OTA by more than half the physical length of any of the other telescopes in the series. Hence, for an 8 inch aperture it is extremely lightweight and portable.

Despite its long focal length, producing a field of view smaller than the SNTs, the SC-8 produces very sharp images of the Planets and deep sky objects. Table 3.5 provides specifications of the LXD SC-8 Model.

Choosing an LXD Telescope

Table 3.4. Schmidt-Newtonian Reflector Specifications

Model	SN-6	SN-8	SN-10
Aperture:	6 inches	8 inches	10 inches
List Price:	$1499.00	$1699.00	$2049.00
Supplied Eyepiece (4000 Series):	SP26	SP26	SP26
Highest Useful Magnification:	254×	271×	339×
Visual Limiting Magnitude:	13.4	14	14.5
Focal Length:	762 mm	812 mm	1016 mm
Focal Ratio:	f/5	f/4	f/4
Resolution (arc seconds):	0.76	0.57	0.46
Net Weight:	48 lbs	69 lbs	85 lbs

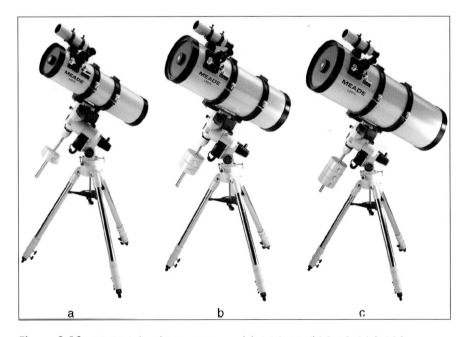

Figure 3.10. LXD55 Schmidt-Newtonian models (a) SN-6, (b) SN-8, (c) SN-10. Image courtesy of Meade Instruments Corp.

A User's Guide to the Meade LXD55 and LXD75 Telescopes

Table 3.5. Schmidt-Cassegrain Specifications

Model	SC-8
Aperture:	8 inches
List Price:	$2399.00
Supplied Eyepiece (4000 Series):	SP2
Highest Useful Magnification:	400×
Visual Limiting Magnitude:	14
Focal Length:	2000 mm
Focal Ratio:	f/10
Resolution (arc seconds):	0.57
Net Weight:	65 lbs

Figure 3.11. LXD55 Schmidt-Cassegrain SC-8 model. Images courtesy of Meade Instruments Corp.

Choosing an LXD Telescope

Features of the LXD Series

Let's review the features that are available with an LXD telescope. Their use will become evident in the chapters that follow.

Optical Tube Assembly Features

The OTA features of the LXD telescope are described below:

Focus Mount. The focus mount is a standard 'Rack and Pinion' design. The drawtube is 2 inches in diameter and comes with a holder which can accept 2 inch or $1\frac{1}{4}$ inch eyepieces. A lock screw adjusts the friction of the drawtube to the focus mount's housing, preventing the drawtube from inadvertently moving when heavy accessories are attached, such as a camera or CCD.

Finderscope. A 6×30 mm achromatic finderscope is used with the N6 reflector whilst an 8×50 mm finderscope is used with the AR, SN and SC ranges. The finderscope is of a different design to other types when it comes to focusing. This is done by loosening the front ring that is located behind the finderscopes' objective lens cell, then screwing the lens cell in or out to focus. When completed the front ring is tightened to lock in the correct focus.

Focus Mount Peripherals. Supplied with the focus mount is a T-ring adapter. This allows equipment such as cameras and CCDs with an M42 screw fitting to be directly connected to the telescope. A diagonal prism is provided with AR and SC ranges. The prism places the viewing position at right angles to the altitude of the telescope, making the viewing of an object in an otherwise awkward position more comfortable, such as objects lying directly overhead.

Dew Shield. A dew cap is supplied with the LXD55/75 refractor range. Made of aluminum, it has a matt black interior to minimize internal light reflections and dew effects. A black plastic dust cap covers the dew shield.

Optical Coatings. To improve the optical quality of the image, the primary lens, corrector plates or mirrors are covered with special coatings. Two types of multi-coatings are used for the LXD55/75 series, EMC and UHTC.

EMC. Enhanced Multi Coatings is used on the Schmidt and Newtonian models (namely SC-8, SN-6, 8, 10 and N-6) in the form of Magnesium Fluoride (MgF_2) coatings for the corrector plates, and standard aluminum on the primary and secondary mirrors. The Refractors use multi-coatings to improve light transmission.

UHTC. Ultra High Transmission Coatings is a set of special coatings Titanium Dioxide (TiO_2), Silicon Dioxide (SiO_2) and Aluminum Oxide (Al_2O_3). When the group coatings are applied to the surfaces of the mirror or corrector plate, it vastly improves the transmission of visible light through them. Hence the brightness of the image by an amount anywhere between 15 and 20% compared to EMC coatings. UHTC is available for the SC-8 Schmidt Cassegrain, SN-6, SN-8, SN-10 Schmidt Newtonian and N-6 Reflector models, but not for the AR Refractor models. UHTC is applied to LXD optics at the factory and cannot be added to telescopes with existing EMC coatings. Therefore UHTC enhanced optics has to be requested at the time of purchase. Many telescope outlets will stock LXD telescopes already with UHTC for an extra $100. It is well worth the additional expense, and I would recommend UHTC to anyone who wishes to purchase an LXD telescope.

Mount Features

The features of the LXD mount are discussed below. Where appropriate, I have highlighted any differences between the LXD55 and LXD75 models. The very first LXD55 mounts had their surfaces coated with a grey finish. Later LXD55 versions had a darker charcoal grey finish and arrow labels were introduced to aid aligning the RA and Dec axes (see Chapter 6). The LXD75 mount however, has a light-grey finish.

Mechanical Features

Motor Drives. Fitted to the mount's axes are two 12V DC. One motor is housed within the control panel plastic housing on the RA axis and the other motor is fitted to the Dec axis. The motors are controlled via controller handsets (see Electronic Features in a later section). Chapter 4 provides a description of how the motors are fitted to the axes of the mount.

Bearing System. The LXD55 bearings are constructed of plastic washers and phosphor-bronze bearings. The LXD75 have stainless steel ball bearings, which are much more hard-wearing.

Polar Alignment Viewfinder. A small telescope has been fitted inside the RA shaft. This is used for polar aligning the mount so that it can track objects in the night sky. The polar alignment scope has become the standard amongst mounts of this type. Attached to the polar alignment viewfinder is a reticle eyepiece, specially designed for polar alignment procedures (Chapter 5). The field of view reveals two etched patterns; a single etched line is displayed for use in the Northern hemisphere, and a four sided polygon represents the circumpolar stars of the constellation Octans in the Southern hemisphere (not the stars around Polaris as stated in the LXD55 manual). The view of the inside of the reticle eyepiece is shown in Chapter 5.

The patterns in the reticle are illuminated via a miniature red LED attached to the side of the polar telescope's eyepiece barrel. The LED is powered by two 1.5 V, button cells (Maxwell LR41 or equivalent). The LXD55 model uses a thumbscrew, which when screwed into the reticle LED holder, illuminates the patterns in the field of view of the alignment scope. The thumbscrew is notorious for accidentally being screwed too far into the holder and making contact with the button cells. If the LED is left illuminated for long periods of time, the button cells power is drained. The thumbscrew can also come completely off with little effort, so is easily lost in the dark if dropped. The LXD75 model improved on this design by replacing the thumbscrew with a twistable switch, which is clicked on and off.

Choosing an LXD Telescope

Latitude and Azimuth Adjusters. The LXD mount uses a design which makes adjustment of the latitude easy and more precise. Two adjustment screws, known as T-handles, use the push-pull lock principle to accurately rotate the telescope's RA shaft around the common centre of the mount. One of the T-handles is loosened whilst the other T-handle applies pressure to the lower part of the mount, and rotates the mount around its centre. This is done until the lower mount has touched the other T-handle. The mount is locked in place with a further tightening of the T-handles. A latitude dial gives the reading of the latitude as the mount is adjusted. Methods for determining the latitude of a particular location on the Earth are described in Chapter 5.

During the polar alignment procedure the mount can be adjusted in azimuth (left-right horizontal movement). This is done using the push-pull lock principle similar to the Latitude adjusters. The adjusters are simple screws that apply pressure to a metal protrusion on the mount section at the top of the tripod. One screw is loosened by adjusting a control knob, whilst the other screw is tightened, causing the entire mount to rotate by a small amount horizontally, to the left or right. The LXD55 model has plastic Azimuth control knobs, whereas the LXD75 model uses aluminum knobs.

Cradle Lock Knob. This is used to lock the cradle assembly arm to the mount. The LXD55 is fitted with a plastic lock knob which encompasses the bolt which locks the arm. If the knob was over tightened, it has been known to 'pop' out the central red plastic covering, which holds the central bolt. The LXD75 lock knob does not have this problem, since the entire bolt and knob is a single cast of solid aluminum.

Manual Setting Circles. Manual setting circle discs are located on each of the equatorial axes. The RA disc displays graduations in Hours and Minutes, whilst the Dec disc displays graduations in Degrees and Minutes. The LXD55/75 series also has digital setting circles which are available through the Autostar system (see Chapter 7).

Counterweights. Counterweights on a telescope are used in order to balance the telescope with the OTA and increase overall stability. The LXD55 series uses a number of counterweights in tandem for each telescope. Each counterweight weighs 10 lbs and has a lock knob to grip secure the weight onto the counterweight shaft. One counterweight is used for the N-6, SN-6 and AR-5. Two counterweights are used for SN-8, AR-6 and SC-8. The SN-10 telescope uses three counterweights.

Electronic Features

The electronic features of the LXD mount are what make the telescope so appealing to astronomers. The control panel allows either a simple electronic controller to be attached, or a sophisticated computerized controller, *Autostar* (see Autostar section).

Control Panel. There are two 8 pin RJ45 sockets, one 4 pin RJ11 socket and a single pin for a 12 V DC power adaptor (Figure 3.12).

Two RJ45 connections provide attachments to the Autostar handset (HBX) and the declination motor (Dec).

An RJ11 socket (AUX) allows accessories to be connected to the telescope, such as a motor focuser or illuminated reticle eyepiece via an accessory port module (see Chapter 12). It has been suggested by Meade, that the AUX port could be used for

Figure 3.12.
LXD control panel.

Meade accessory products that will be available in the future, although at the moment there is very little that will connect to it.

The power socket accepts a standard power supply plug from a 12 V DC transformer or battery pack. The battery pack holds eight 1.5 V D-cell batteries and is expected to last 30+ hours. A 12 V car or leisure battery can also be used as a power source.

When power is supplied a red LED illuminates indicating that the switch is in the 'on' position.

Autostar

Autostar has been available for several years and is the standard handset used among owners of Meade telescopes. There are various different models available including Autostar II, #494 and #497. The LXD series uses the #497 model (Figure 3.13).

Autostar is multi-functional; it is more than just an electronic motor to slew and track the telescope. It *is* the Goto system, and without it the LXD would be just another equatorial telescope series, competing with other telescope brands. The handset is supplied as standard with all models apart from the N-6 Reflector for which it is optional. Hence the handset is available separately for around $149.

The features of the Autostar handset are described throughout the book and will not be discussed here. The handset contains 30,223 objects (see Chapter 8 for a detailed breakdown). You can store up to 200 of your own user-defined objects, and also use it to calculate astronomical events such as Solar and Lunar Eclipses, Sun and Moon rises as well as equinoxes for any date provided. It also includes a limited glossary of astronomical terms. Functions are found via simple navigational menus using up and down menu keys located at the bottom left and right sides of the handset.

Autostar software can be upgraded from a computer via a Serial interface (Chapter 10). Software upgradeability is an advantage as it means that the latest objects information such as Comet and Satellite data can be up-linked to the handset as well as bug fixes to improve operational efficiency. It can even function as a red LED torch!

Armed with an Autostar and its powerful Goto feature, you will be able to explore Universe, all with just a few keystrokes.

Choosing an LXD Telescope

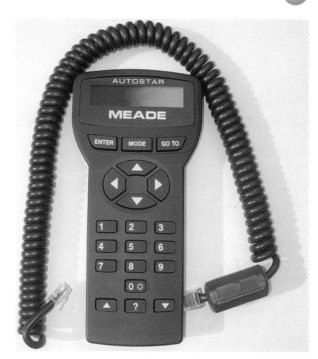

Figure 3.13. Autostar computer controller.

Autostar Specification Summary:

Handset Specs:
- 20, button alphanumeric keypad
- 2 line 16 character LED display
- Keypad Backlight (red led)
- 4 pin RJ11 Serial RS232 Socket
- 8 pin RJ45 Socket to connect to HBX on control panel
- Weight 1.12lbs
- Price $149

Internal Component Specs:
- Motorola 68HC11 Chip speed rated 8MHz
- PIC 16C57 Microcontroller
- 1.0Mb Reloadable Flash Memory

The LXD Tripod

For both LXD tripods, the height is adjustable. When fully extended the top base of the tripod is high enough above the ground for comfortable viewing with any of the telescopes in the series. The whole tripod assembly can be opened up and locked in less than a minute.

LXD55. The tripod is made of rectangular hollow struts constructed of lightweight Aluminum. There are two per leg. The struts are connected to a top base via a single wing bolt and the telescope mount head is attached to the tripod through an M10 screw thread and mount locking knob. A star symbol ☆ is embossed on top of one of the legs, which represents the direction of a polar region, North or South.

Attached to the three leg braces of the tripod is a plastic accessory tray. The tray clips in place onto the leg braces and a single small plastic locking screw-cap is used to hold the tray in the centre where the three leg braces meet. A rubber tip is used at the bottom of each leg to reduce vibrations of the telescope.

LXD75. The LXD75 tripod has been completely re-designed and is much superior to its LXD55 predecessor.

A new heavy-duty stainless steel tripod makes use of a locking spreader bar to lock each leg in place. This almost eliminates the flexure suffered by the LXD55 tripod. The spreader bar also has holders for three 1.25" eyepieces. Each leg has two aluminum leg extension bolts which when locked prevents the internal leg from slipping when extended. The locks face inwards, so that a person does not accidentally kick them in the dark and knock the mount out of polar position, or cause injury.

Rubber tips are also used at the bottom of each leg like the LXD55 tripod to reduce telescope vibrations.

Conclusion

Table 3.6 provides a summary of LXD telescopes, to show which are more suited for a particular type of observation. This acts a guide only, since all telescope models will perform satisfactorily on the objects listed.

I have highlighted in this chapter many different types of instruments that are on offer, and have provided reasons why you should choose a telescope that is best suited for your particular interest in astronomy. There is no hard and fast rule for choosing which telescope to purchase. In the end it could depend on how much you want to spend, or whether you want a telescope to do most of the work for you.

'Caveat Emptor' is Latin for 'Let the buyer beware!' This holds very true when purchasing a telescope either for the first time or for the experienced buyer. It is comforting to know that the numbers of con dealers or adverts in the newspaper selling tiny telescopes, which boast unrealistic super magnifications are dwindling. However,

Table 3.6. Observations Type Versus Telescope Model

Observation Type	Telescope Model
Solar System	N-6, AR-5, AR-6, SC-8
Stars	N-6, AR-5, AR-6, SC-8
Deep Sky	SN-8, SN-10, SC-8
Imaging (Traditional and Digital)	SN-6, SN-8, SN-10, SC-8

Choosing an LXD Telescope

dealers of this type still exist, so beware, as you will always end up disappointed with what you buy from them.

Thankfully, most telescope dealers nowadays provide sound advice for beginners on what instrument they should purchase as their first telescope. The best way to find out the reputation of some dealers is to attend a local astronomical society and meet other astronomers who have experience in purchasing instruments. They will best advise you where to purchase and what the current best deal is, so that you get value for money.

As an owner of two Meade LXD models, I would unashamedly recommend this series of telescopes (that is what this book is about after all!). Inevitably, whatever you decide to buy, you will have embarked upon the most exciting hobby there is around.

CHAPTER FOUR

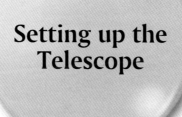

Setting up the Telescope

Setting up any kind of new instrument for the inexperienced can be stressful and frustrating at the best of times. Anyone who has ever purchased a DVD player or VCR will no doubt have had a frustrating experience when trying to set it up, using the instruction manual that comes with it.

It is not my intention to suggest in anyway that this chapter should replace the manual that is shipped with the telescope, far from it! The manual is adequate enough to help those new to setting up a telescope and get it operational. However, in order to help you set up a telescope with relative ease, I intend to re-examine some of the key set up tasks and provide a description of setting up the LXD55 and LXD75 telescopes, from my own personal experience.

In some places you may find this chapter a bit formal to read and very manual-like. This is purposely done as it is intended to act primarily as a reference, where relevant sections are read independently rather than the whole chapter page to page. I have tried however, to place the sections in a particular order as you would carry them out from the time the telescope is first taken out of its box and assembled.

Most of the set up procedures in this chapter are common for both LXD55 and LXD75 telescopes. If a procedure is intended for just one of the telescope series, then it will be highlighted for that case. All times are conservative estimates, based upon how long it would take someone experienced to perform the setup procedure plus additional time for someone who is less familiar with setting up a telescope.

I would suggest that you read all the steps for each procedure first before attempting to carry them out. Some procedures require the use of the Autostar, so you need to be familiar with its menus (Appendix C). Once the telescope has been set up, you should fully familiarize yourself with its operation in the daylight before taking it out for the first time at night, so it won't be a disappointing experience should things go wrong. A good knowledge of telescope operations is the key to user confidence!

Setup Tasks

Assembling the Telescope (ECT: 2 to 3 Hours)

Personally, I find one of the most exciting moments once I have purchased a telescope is when it has finally been delivered, and it is sitting there in its packaging on my living room floor ready to be opened. The waiting for the delivery is over, and I get excited hoping that the instrument I am about to assemble will provide me with many years of enjoyment. Who knows! I might even discover something with it one day!

The telescope is packed tightly in foam in large boxes in order to safely transport the fragile instrument to its final destination. Damage can occur to the instrument sometimes during transit, so you should check the packaging thoroughly for signs of external damage or tampering before opening the box. It is not recommended to dispose of the packaging immediately (unless you have problems with storage) just in case there is a problem with the telescope and you need to send it back under warranty.

Assembling the telescope should be straightforward but not a trial. It shouldn't take you longer than an afternoon (or evening) to unpack and assemble all the relevant components together (Figure 4.1).

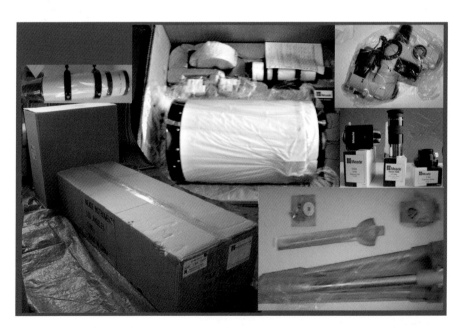

Figure 4.1. Telescope components straight out of the box.

Setting up the Telescope

Handling the Telescope

Some telescopes are heavy and bulky to handle. As an owner of a telescope you will be constantly assembling or taking apart the telescope, so you will have to take special care every time to make sure you don't damage the instrument as well as injure yourself every time you pick up the telescope and move it around. It is a fact that most of the damage done to a telescope is during assembly where the telescope and associated components are not assembled in the proper manner. You should also consider assembling the telescope in a 'clean' area where there is little clutter to get in the way.

Telescopes are purposely split into three main components (OTA, Mount, Tripod stand) in order to make them as portable and easy to handle as possible. There is a right and a wrong way in lifting the three components when assembling or packing the telescope away for storage. I don't claim to be a strong person myself, and to be honest I have sometimes struggled with lifting long telescope tubes onto heavy mounts, as well as trying to move the entire telescope assembly around, which I don't attempt anymore.

Handling the tripod is straight forward. It is best to hold two of the three tripod legs with both hands when you lift the tripod to carry it around (Figure 4.2). If you hold just one of the legs with one hand that the tripod is likely to open up unexpectedly, as you are carrying it and you may trip over.

The mount and the counterweights are the next heaviest to handle and carry around. I tend not to have the counterweights already attached but attach them to the mount after it has been secured to the top of the tripod. Both hands should be used to lift the mount. One hand should be placed near the counterweight shaft screw cavity, whilst the other hand should be placed just under the polar finderscope housing (Figure 4.3). Do not hold the mount by either of the motor housings, as you may pull them slightly away from their shafts, especially if the bolts holding the motors in place have not been sufficiently tightened.

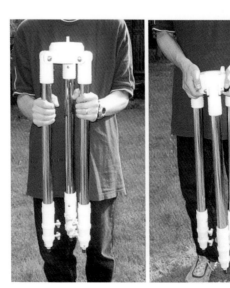

Figure 4.2. Holding the LXD tripod stand.

Figure 4.3. Holding the LXD mount.

Handling the OTA requires the greatest of care due to its fragile elements. Refractors and Reflectors OTAs are quite bulky due to their long lengths, whilst the SCT has a very short tube length design and is the easiest to handle. You will need to hold the OTA correctly when you either place it onto the mount or take it off. Both the LXD55 and LXD75 refractors and SCT OTAs have chrome handles attached to them to help users when they are lifted whereas the SNTs and the 6 inch reflector lacks them as part of their assemblage. If you have any doubt at all about handling the OTA on your own, you should get a second person to help you.

Lifting an OTA should be done with extreme care and plenty of patience. When you pick up the refractor you should grab the chrome handle with one hand and then place your other arm underneath the tube at a position that is the most comfortable for you. Do not hold the refractor by the dew shield, as it will come away in your hand. Try to keep the OTA as close to your chest as you can when you are carrying it around. The 'fun' part is trying to place the OTA onto the mount. If it is already attached to the cradle assembly, which is often the case, try to make sure that the cradle assembly is facing down towards the ground, to make it easier for you to place the OTA into the slot on the top of the mount, in one single action. You may wish to set up the mount and align it with the celestial pole before attaching the OTA.

Sometimes, it does feel like you need three hands to attach the OTA to the mount, two hands to hold the mount and that imaginary third hand to tighten the lock knob to hold the cradle assembly shaft to the mount. In reality, using just the two pairs of hands, you will at some point need to let go of the OTA with one hand, in order to tighten the lock knob on the side of the mount slot. You should first slot the cradle shaft

Setting up the Telescope

Figure 4.4. Handling the refractor OTA.

into the mount slot then let the telescope rest against your upper torso or shoulders whilst a free hand tightens the lock knob. You should be able to allow most of the weight of the OTA to be taken up by the mount. This can be quite a risky task to carry out and you will have to be extremely careful that you do not give yourself an injury, or damage the instrument, whilst you are precariously balancing the OTA against yourself. Once again, if in doubt don't do it yourself, get an extra helping hand. Figure 4.4 shows how to handle the Refractor OTA.

The SNTs are almost as long as the refractors however since they don't have chrome handles to assist you in lifting them, one hand should be placed under the mirror cell whilst the other hand is placed further up the OTA at a position that is comfortable to hold, i.e. equidistant from the common centre of gravity of the instrument.

Handling the SCT OTA is easier than the rest of the other telescopes due to its compact design, lightweight construction and a chrome handle. You will still need to make sure that the OTA mount shaft is facing down, and you will still need to take care and carefully poise the OTA, whilst you are locking the shaft into the mount slot (Figure 4.5).

Figure 4.5. Handling the SCT OTA.

Removing the OTAs from the mount also requires care and thought. The task is done in reverse order to that which has been described earlier in this section. Remember to make sure that you have your torso rested against the larger OTAs, before you undo the lock knob for the mount slot.

Balancing the Telescope

Every astronomical telescope needs to be balanced. The aim of balancing a telescope is to ensure smooth operation, enhance pointing accuracy, increase the motors operational life and increase the overall telescope's stability. Once a telescope has been precisely balanced, you should be able push the telescope into any position in either the RA or Dec axes without moving on its own accord and without any tension from the locking knobs.

The telescope should be balanced in both axes. Preferably, you should balance the telescope with accessories that would be used during normal operation, such as the 26 mm eyepiece that is supplied as standard. Other accessories such as cameras, CCDs or large heavy eyepieces will likely require re-adjusting the telescope's balance.

You should periodically re-balance the telescope to keep it operating at peak efficiency.

Balancing the AR Refractor, SN Schmidt-Newtonian and N-6 Newtonian Telescopes (ECT: 15 to 25 Minutes)

Before starting the balance procedure, ensure that the middle section of the OTA lies between the two telescope slip rings or telescope cradle. This will require undoing the slip ring screws and sliding the OTA through the rings or cradle until it is suitably positioned. Make sure that these screws are sufficiently tight when finished. Check that the counterweights are firmly fixed on the counterweight shaft (somewhere halfway along the shaft), and that the counterweight safety cap is screwed on tightly (so as to not break your toes should a counterweight slide off the shaft!). Ensure that the 26 mm eyepiece is in the focus mount.

Setting up the Telescope

Figure 4.6.
Unbalanced AR-6 OTA in Dec axis.

Dec Balancing.

1. Make sure that the RA axis is locked in place. With one hand on the OTA loosen the Dec lock.
2. If the telescope is out of balance it will rotate around the Dec axis on its own accord. Once again, one hand should be gently rested on the OTA, to prevent any damage in case the OTA rotates too rapidly. Figure 4.6 shows an example of the AR-6 OTA movement when it is out of balance.
3. Note which way the OTA rotates around the Dec axis. If the *lower* half of the OTA rotates towards the ground (denoted by 'a' in Figure 4.6) then the whole tube assembly needs to be moved further *up* through the cradle rings (shown as up arrow in Figure 4.7). Alternatively, if the *upper* half rotates towards the ground (denoted by 'b' in Figure 4.6) then the OTA needs to be moved further *down* through the cradle rings (shown as down arrow in Figure 4.7). Remember to lock the Dec each time before moving the OTA up or down through the cradle rings.

Figure 4.7. OTA movement through cradle rings.

Figure 4.8. OTA with counterweight shaft horizontal with the ground.

A balance in the Dec axis is achieved when the Dec axis is unlocked and the OTA stays in position.

RA Balancing. To balance the RA axis, the counterweights needs to be in the right position along the counterweight shaft:

1. With one hand on the OTA loosen the RA lock. Make sure that the Dec axis is locked in place.
2. If the telescope is significantly out of balance it will rotate around the RA axis under its own accord. One hand should be gently rested on the OTA in order to prevent any damage or injury in case the OTA rotates too rapidly.
3. Orientate the OTA, so that it is horizontal in parallel with the ground as shown in Figure 4.8.
4. Note which way the OTA moves. If it moves upwards then the counterweights on the opposite side needs to be moved **up** the shaft. If the OTA moves downwards then the counterweights need to be moved **down** the shaft. Remember to temporarily lock the RA when moving the counterweights up or down the shaft. Then unlock it when you are ready to check the balance of the OTA again.

A balance is achieved when the RA is unlocked and the OTA stays in position.

Balancing the SC-8 Schmidt-Cassegrain Telescope (ECT: 10 to 15 Minutes)

The method for RA balancing an SCT is identical to the LXD55 and LXD75 telescopes.

There are no cradle rings for an SCT, so the method for balancing the Dec axis is slightly different to that of other telescope OTAs; the entire SCT OTA is moved up or down along the cradle assembly slot, rather than through cradle rings (Figure 4.9).

Setting up the Telescope

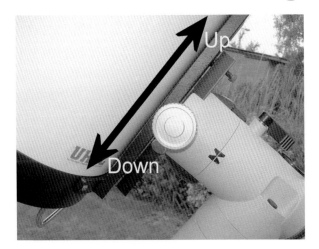

Figure 4.9. SCT OTA movement along cradle assembly slot.

Polar Viewfinder Alignment

To accurately polar align the mount with the North Celestial Pole the polar viewfinder has to be in perfect alignment with the RA shaft. It is common for telescope mounts with built-in polar alignment viewfinders, to find that the viewfinder is not completely aligned inside the RA shaft, when the telescope is shipped straight from the factory. The manual supplied with the LXD series does not describe any methods for aligning the polar viewfinder within the RA shaft, as it assumes they are roughly aligned, which is correct in most cases.

It is very important to have the viewfinder physically aligned with the RA shaft. Misalignment can lead to problems with tracking objects and Goto inaccuracies.

The view through the polar viewfinder is shown in Figure 4.10.

The vertical line with gradations is used to align the telescope for the northern hemisphere whereas the quadrangle shape is used to align the telescope for the southern hemisphere. I will discuss how to use these shapes to polar align the telescope in Chapter 5.

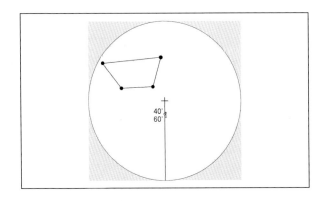

Figure 4.10. Schematic view through the polar viewfinder. Image courtesy of Alan Marriott.

The alignment of the polar viewfinder with the RA shaft is a two stage process. The first stage determines whether the polar viewfinder is perfectly aligned with the RA shaft and the second stage is the physical adjustment of the viewfinder.

Polar Viewfinder Shaft Alignment Test (ECT: 10 Minutes)

It usually becomes awkward to view the object with the OTA attached because of the tripod legs getting in the way of the attachments, so it is advisable to remove the OTA and the counterweights before you carry out this test.

1. Set the latitude of the mount, so that the polar viewfinder is horizontal with the ground (Figure 4.11).
2. Point and centre the polar viewfinder's central crosshair on a fixed terrestrial object such as a distance tree or landmark at least 100 meters away. You will notice that objects through the polar viewfinder appear upside down.
   Note: The stars that are used for polar alignment are essentially at infinite distances away hence, the further away the terrestrial object the more accurate the alignment will be. Moreover, the procedure works best if the polar viewfinder is horizontal with the ground. However, if a suitable object cannot be found in the horizontal position, then the latitude of the mount can be adjusted, until an object is seen in the polar viewfinder. This will not affect the final outcome of the procedure.
3. Unlock the RA axis and rotate the telescope until it is in position 1 (Figure 4.12a). Keep looking at the distant object in the polar viewfinder whilst the RA shaft is being slowly rotated into position.
4. Now, slowly rotate the telescope until it is in position 2 (Figure 4.12b), whilst again looking through the polar viewfinder at the object.
5. Note the movement of the object in the polar viewfinder whilst the telescope is rotated around the RA shaft. If the central crosshair in the polar viewfinder stays

Figure 4.11.
LXD mount set in horizontal position.

Setting up the Telescope

Figure 4.12. (a) OTA position 1 and (b) position 2.

exactly on the object's centre, or very close to it, then the polar viewfinder is aligned with the RA shaft and no further action needs to be taken. If the object moves in any significant direction whilst the telescope is being rotated around the RA shaft then the polar viewfinder will require aligning with the RA shaft, and the second part of the alignment process should be carried out.

Aligning the Polar Viewfinder With the RA Shaft

(ECT: 30 to 45 Minutes)

There is a tried and tested method which is used to align the polar viewfinder with the RA shaft. This part of the procedure can be tricky to understand, so carefully refer to the examples shown in Figure 4.14a–f. Basically the method is aimed at reducing the misalignment error between the viewfinder and the RA Shaft with each adjustment until the error is eliminated.

To start, you need to locate three adjustment grub screws at 120° positions around polar viewfinder barrel (Figure 4.13). These three screws hold the polar viewfinder's eyepiece barrel in place and will be used to adjust the position of the barrel in the RA shaft.

6. Rotate the RA shaft until it is in position 1 (Figure 4.12a). Centre the object onto the central crosshair as shown in Figure 4.14a. Make sure you centre onto an obvious feature such as a church steeple or a kink in a branch on a tree. If you are using trees to do the alignment, you should do it on a calm day as the swaying of the trees will stop you from performing the procedure successfully.
7. Now, rotate the RA shaft until it is in position 2 (Figure 4.12b). The view is shown in Figure 4.14b. Note the movement of the object as it moves away from the central crosshair. This is a sign of misalignment and you should make a mental note of the object's distance to the original position, when the RA shaft was in position 1.

A User's Guide to the Meade LXD55 and LXD75 Telescopes

Figure 4.13. Polar viewfinder and grub screws (arrowed). Note: Third grub screw not pictured.

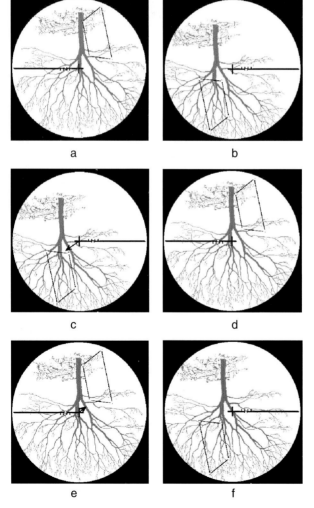

Figure 4.14. (a to f) Schematic images of a tree seen through the polar viewfinder.

Setting up the Telescope

8. Carefully undo the adjustment screws and adjust the eyepiece barrel of the finderscope in the RA shaft until the central crosshair is positioned almost HALFWAY between the centre of the object at its current position and the original position of the central crosshair when the RA shaft was first rotated (Figure 4.14c).

   The positioning of the eyepiece assembly in the RA shaft is actually quite a tricky task. You may find it difficult keeping the eyepiece in that halfway position with one hand, whist you tighten the three adjustment screws using an Allen key with the other. It takes practice and patience to get it right, so take your time.

9. Rotate the RA shaft back the other way until it is in position 1 again (Figure 4.14d). If the object is still off-centre then repeat steps 6 thru 8 (Figures 4.14e and 4.14f) continually readjusting the eyepiece barrel until the object stays relatively near the central crosshair when the telescope RA shaft is rotated between position 1 and position 2. A near-perfect centralized alignment of the viewfinder can be achieved using this method. This will be more than sufficient to carry out polar alignment of the RA axis.

You will need to carry out alignment of the polar viewfinder with the RA shaft usually just the once over the lifetime of the telescope. The exception is if the viewfinder barrel is taken out for cleaning or replaced.

The Polar Home Position (ECT: 5 minutes)

For most telescopes with a Goto facility there is a default position, where the telescope initially starts from before Goto operations can begin. This is known as the Polar Home Position.

The telescope is deemed to be at the Polar Home Position if the following criteria are met:

1. The tripod's north leg (marked with a star if LXD55) is pointing North.
2. The latitude of the mount is set and pointing to the North Pole Star Polaris (see polar alignment Chapter 5). The same principle applies for the South Celestial Pole.
3. The Dec axis is rotated such that the OTA is exactly at right angles to the counterweight shaft.
4. The RA axis is rotated such that the counterweight shaft is pointing straight down over the mount.
5. The two arrow markers on the side of each axis are aligned. (It is possible that these have been misaligned during manufacture, which in this case the arrows will need re-positioning. See Chapter 13 for troubleshooting).

The Autostar has a feature which allows you to 'Park' the telescope when a nights observing session is over. The final resting position is the Polar Home Position (Figure 4.15). I will discuss more about the benefits of 'Parking' the telescope in Chapter 6.

Figure 4.15. The Polar Home Position.

Aligning the OTA With the RA and Dec axes

In order to obtain good pointing accuracy during tracking and Goto operations, the OTA needs to be physically aligned in parallel with the RA axis. Also the Dec axis needs to be calibrated with the Dec setting circle in order to assist with other setup procedures, and to provide accurate Declination co-ordinates. Once the RA axis is aligned with the OTA, all objects in the eyepiece should rotate about a central point, without any type of image shift in any direction (up, down, left or right).

The methods described are similar in principle to aligning the polar viewfinder with the RA shaft as described earlier in this chapter, i.e. the reduction of misalignment errors through repetitive adjustment procedures. There are a number of methods to carry out and you can decide which method works best for you. You may find these methods are overkill if you just want to go out and do some basic observing. Using the telescope straight out-of-the-box is normally good enough to perform simple observing tasks. However, if you want to make sure that the Goto and tracking features will work to maximum accuracy and reduce 'cone error' (see Chapter 5), then I would strongly advise you to carry out the alignment procedures outlined in this chapter.

Setting up the Telescope

The Dec axis calibration has to be carried out before the RA axis alignment procedure is performed. The Autostar will be used to move the telescope during these procedures and, so you should be familiar with basic Autostar operations.

Dec Axis Calibration (ECT: 15 to 20 Minutes)

The result of calibrating the Dec axis is to set up the Polar Home Position of the telescope as described earlier in the chapter.

The procedure is as follows:

1. Power on the Autostar handset.
2. Set the telescope to position 1 as shown in Figure 4.12a.
3. For Schmidt-Newtonian models you should ensure that the OTA is rotated in the cradle rings, such that the focuser is horizontally parallel with the ground whilst in the position 1 stance. You should NOT rotate the OTA during this procedure.
4. If using the diagonal prism for Refractor models ensure that it is set flush against the focuser at all times.
5. The shaft upon which the cradle rings assembly is attached to should be set flush against the sides of the slot on top of the mount. The cradle lock and smaller silver knobs that lock the shaft in place should be tight.
6. Whilst in position 1, point the telescope to a distant horizontal target by adjusting the polar altitude/azimuth and tripod height, but NOT the RA or Dec axis. Both axes must stay locked.
7. Centre the target using the 26 mm eyepiece supplied with the telescope or one that is slightly higher in magnification such as a 20 mm or 15 mm eyepiece. If there is a problem with the counterweight shaft knocking into one of the tripod legs then adjust the length of the back tripod legs until this no longer happens.
8. You should make sure that the Dec axis is locked. Rotate the RA axis, so that it is in position 2.
9. If the target image has shifted up or down in the eyepiece (ignore left-right shifts for the moment) then use the Autostar up/down arrows keys (use an appropriate slew speed) to move the target image until it is HALFWAY back to the centre of the eyepiece.
10. Rotate the telescope back to position 1 and note if there is any up/down shift again. If there is a vertical shift of the image in the eyepiece, then adjust the position of the image again as in step 6 using the Autostar up/down arrows.
11. Repeat steps 8 thru 10 until there is no vertical shift in the image when the telescope is moved from position 1 to position 2.
12. To complete the procedure, set the Dec setting circle so that it reads 90° whilst the telescope is in position 1 or 2. This means that when the telescope OTA is positioned above the RA shaft, it will correspond to a Dec position of 0° (the Polar Home Position).

Care needs must be taken to ensure that the Dec setting circle is not accidentally shifted, otherwise and the setting circle will provide inaccurate readings.

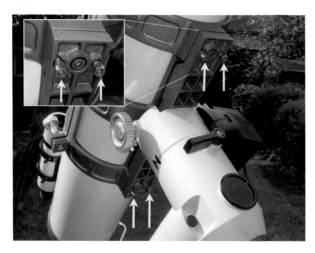

Figure 4.16. Cradle adjustment screws (arrowed) and close up (inset).

RA Axis Alignment

Now that the Dec axis has been calibrated the OTA can be aligned with the RA axis. The adjustment of the OTA in RA is done using three hexagonal screws located under the cradle assembly shaft (Figure 4.16).

Note: Make sure that the Dec axis is not moved at any point during the RA alignment procedure.

Method 1 (ECT: 10 to 15 Minutes).

1. Rotate the telescope around the RA shaft until it is in position 1 (Figure 4.12a).
2. Point to a distant target by adjusting the tripod and the polar azimuth and latitude. Centre the target in the eyepiece.
3. Rotate the tube in RA only until it is in position 2 (Figure 4.12b).
4. Note the position of the target in the eyepiece. If the RA axis has not been aligned then the target's position will have shifted sideways to the left or right.
5. Using both sets of cradle rings RA adjustment hexagonal screws, move the target until it is HALFWAY between the current position and the centre of the eyepiece. This is done by undoing the outer two hexagonal screws and then turning the central hexagonal screw to move the OTA in latitude by a small amount, until the target is in the halfway position. When finished tighten the two outer screws. You may be constantly tightening and loosening the screws, so you may need some patience to get it right.
6. Now again centre the object in the eyepiece by physically adjusting the polar azimuth and latitude and tripod legs only (the Dec axis on the mount should not be adjusted in any way).
7. Repeat rotating the tube around the RA shaft until it is in position 1 again.
8. Repeat steps 3 thru 7 until the target image moves not at all or only very slightly when alternating between position 1 and 2. You can also try a different target to

Setting up the Telescope

test the alignment if you wish to double check that the alignment procedure was successful.

Method 2 – Autostar LXD55/75 Adjust Feature: (ECT: 10 to 15 Minutes). The 'LXD55/75 Adjust' Autostar feature can be used to align the OTA with the RA axis. The only disadvantage to this method is that the settings will be lost if a full 'Reset' of the handset is carried out and the whole procedure will have to be performed again. There are some points to note before you start the alignment procedure:

- The RA shaft needs to be set to more than 45° in latitude.
- Ensure than the telescope can be rotated 180° in both RA and Dec axes without obstruction.
- Ensure that the telescope is rotated in Dec such that the Dec motor is positioned over the RA shaft.
- Set the OTA, so that it is roughly horizontally parallel with the ground.
- For Schmidt-Newtonian models you need to ensure that the focuser is pointing in parallel with the RA shaft.
- For Refractor models you need to ensure that the diagonal prism is set flush against the focuser.
- The telescope motors should have already been trained and calibrated.
- Ensure that both RA and Dec axes are locked.
  1. Switch on the Autostar and go to the 'LXD55/75 Adjust' option. (This can be found by scrolling to Setup→Telescope→LDX55/75 Adjust).
  2. When pressing 'Enter' on the 'LXD55/75 Adjust' option a message asking to centre a landmark will be displayed.
  3. Centre a landmark in the eyepiece. This is done by moving the entire telescope/tripod assembly until the landmark is in the field of view of the eyepiece. You should not centre the landmark using the RA and Dec axis. These should be locked in position and must not be adjusted.
  4. Press 'Enter' on the Autostar once the landmark has been centered.
  5. Another message is displayed asking to repeat centering the Landmark in the eyepiece. Pressing 'Enter' will start the rotation of the telescope OTA in both axes (first in Dec 180° then in RA 180°) with the message 'Slewing...' displayed on the Autostar. When this is happening, you should stand back from the telescope and make sure that the OTA will not hit anything while it's rotating around the axes. If a problems occurs whilst the OTA is in the middle of a rotating motion, i.e., the OTA catches on something like a motor housing, then you should immediately power off the telescope and repeat this procedure again. Don't worry! You will not damage the Autostar or the telescope as a result of powering off the telescope whilst the OTA is in mid-rotating.
  6. When the rotation has completed, the Autostar once again asks you to centre a landmark in the eyepiece. It should be the same landmark as the one first chosen. This time use the arrow keys on the Autostar to centre the landmark.

7. Pressing 'Enter' again will rotate the RA axis and when it has stopped centre the landmark in the eyepiece, this time though using both sets of cradle rings RA adjustment hexagonal screws. The OTA should now be aligned with the RA axis.

Method 3 – Using the Polar Viewfinder (ECT: 10 to 15 Minutes). Another way to align the OTA with the RA axis is using the polar viewfinder situated in the RA shaft. At a Dec position 0°, the OTA will be directly above the RA shaft. If they are aligned, the image you see through the polar viewfinder will also be in the eyepiece of the OTA.

You should have carried out the following two procedures in order to perform this method of alignment:

- The viewfinder should be centered in the RA shaft. (See the Polar viewfinder alignment procedure earlier in this chapter).
- The Dec axis should be calibrated such that at 0° the OTA is almost parallel with the RA shaft (the Polar Home Position).
  1. Set up the telescope, so that it is in position 1 (Figure 4.12a). The Dec axis should be at the Polar Home Position and locked in place.
  2. Centre an object onto the central cross hair in the polar viewfinder. This is done by physically moving the whole telescope/tripod assembly. Note that it does not necessarily have to be an object directly on the horizon, so the latitude of the mount can also be adjusted in order to locate an appropriate object.
  3. Note the position of the object in the eyepiece of the OTA (use a low power eyepiece such as the 26 mm). If the Dec axis has been set to 90° then the object should be in the field of view in the eyepiece.
  4. Adjust both sets of cradle rings RA adjustment hexagonal screws until the object in the OTA field of view is HALFWAY to the centre of the eyepiece.
  5. Now rotate the telescope until it is in position 2 as shown in Figure 4.12b, whilst checking that the object is still centered in the polar viewfinder. Again note the object in the OTA eyepiece field of view and repeat step 4.
  6. Repeat the entire procedure again until the object is in both the centre of the OTA and the polar viewfinder in both positions 1 and 2.

If the object is not completely centered in the eyepiece when moving from positions 1 to 2, then it means that the Dec axis has not been perfectly calibrated to 0°.

The LXD Motor Drives

For telescopes fitted with motor drives there are always imperfections when the motors are in operation. These imperfections can lead to inaccuracies in Goto and tracking operations. The Autostar has a number of utilities which are used in order to reduce these inaccuracies as much as possible:

- Calibrating the motors.
- Backlash compensation.
- Training the motors.

Setting up the Telescope

Once again you should be familiar with Autostar operation before using the utilities listed above. It should also be noted that these utilities will have to be repeated if a full reset of the Autostar handset is carried out.

Calibrating the Motors (ECT: 5 Minutes)

This utility is usually carried out if a problem develops with the motors, such as unusually noisy operation or there is an erratic problem with driving the telescope in either the RA or Dec axes.

The motor calibration utility can be found on the Autostar by scrolling to the option below. Press 'Enter' to start the calibration procedure.

Setup→Telescope→Calibrate Motor

The utility is often carried out before a full Reset is performed, but there are no ill-effects if the motors are calibrated during the normal operation of the telescope. In fact, the motors should be calibrated periodically (monthly) in order to increase their lifetime, and to improve the accuracy of the Goto operations.

If after calibration there are still problems with the motor drives, then you may have to contact Meade as it is possible that the motors are at fault. (See Chapter 13 for troubleshooting the motors).

Backlash

Most telescope mounts including the LXD75 and LXD55 series employ the worm-gear and wheel assembly design. The worm is a shaft made out of aluminium with a gear thread moulded on to it. The wheel is designed such, that the worm-gear thread fits comfortably into a threaded groove on the outside edge of the wheel.

Backlash occurs when the worm and wheel do not perfectly align with one another. There is a small amount of 'play' between the worm thread and wheel, such that a small incremental turn of the worm is required, before the teeth engages into the wheel's groove and rotates it.

Backlash is one of the most common causes of tracking problems on motor driven telescope mounts. For the worm and wheel arrangement, backlash cannot be fully eliminated, even if the mount has been expensively engineered. Fortunately telescope manufacturers nowadays provide electronic hand controllers including the Autostar, which incorporate a backlash compensation utility. However, some telescope mounts, which come straight from the manufacturer have the worm and wheel arrangement so badly misaligned that problems, such as the inability to track objects and severe Goto errors often occurs. In these cases, electronic backlash compensation will not solve the problem, and warrants a physical adjustment of the worm thread and wheel assembly instead. In such cases, the compensation utility will not completely eliminate the backlash of a telescope mount, but it will significantly reduce it to a satisfactory operational level.

Autostar Backlash Compensation Utility (ECT: 15 to 25 Minutes)

On the Autostar, the backlash compensation utilities for RA and Dec axes can be found by scrolling to the following options:

| Setup→Telescope→Az/RA percent |

| Setup→Telescope→Alt/Dec percent |

The percentage number entered determines the speed of the motor response, when the Autostar navigation arrow keys are pressed. A value of 100, implies that the worm thread that the motor is attached to, will respond almost immediately when the arrow keys are pressed. The wheel fixed on one axis will rotate as a result of the worm turning it and hence, the telescope will react more quickly. A value of 0 implies a longer response in the movement of the worm to the wheel on the axis. So the telescope's reaction will be slower. Backlash for each new telescope of the LXD series is slightly different. You will have to experiment with different values until you find the correct response for your telescope and motor system. It is often the case that RA and Dec percentage values are not the same.

Physical Motor Alignment (ECT: 45 to 60 Minutes)

Physical re-alignment of the worm-wheel assembly requires loosening the motors from the mount, and adjusting their position slightly until the 'play' between the worm and wheel is sufficient enough, so that the backlash compensation utility on the Autostar handset can be effective.

Important Note: You must be careful that you do not push the motor unit too far against the groove in the wheel. You might cause damage to the motor itself. Over-tightening of the worm thread to the wheel will mean that the motor will struggle to turn the worm, due to excessive binding and therefore could cause damage to the motor's internal workings.

RA and Dec Motor Unit Alignment

The Dec Motor Alignment should be done first as it is relatively easier and quicker to carry out than the RA Motor alignment. The hexagonal bolts used to attach the Dec motor to the mount are more accessible, than for the RA motor unit (Figure 4.17).

The RA motor unit is attached in a similar way as the Dec motor. The only difference is that the location of the RA motor unit main bolt is in a more awkward position than the Dec one. The bolt is located under the polar alignment telescope, and you have to adjust the latitude of the mount to an almost vertical position, so that you can get a clear view of the bolt (Figure 4.18).

Setting up the Telescope

Figure 4.17. Dec motor housing with hexagonal bolts (arrowed).

Figure 4.18. RA motor housing with hexagonal bolt (arrowed).

Figure 4.19. Drive cogs in the motor housing.

The steps to align the RA motor unit are similar to the Dec motor unit alignment. The difference is that you use the Left and Right arrows on the Autostar handset for RA instead of the Up and Down arrows for DEC.

For the RA motor alignment procedure, remove the OTA and adjust the latitude so that the RA axis is almost horizontal with the ground.

1. Remove the side plastic panel on the motor unit to reveal the inner drive cogs (Figure 4.19).
2. Undo the two hexagonal screws either side of the motor unit as shown in the diagram. You only need to loosen the bolts rather than remove them from the mount altogether.
3. Loosen the central bolt such that the whole motor unit is free to move about.
4. Gently push the motor unit you are adjusting towards the Dec or RA axis until you can sense that the worm and wheel are against one another, but not so tight that they bind together.
5. Look at the cogs on the side of the motor unit and determine if they look as if they are coupled together. I.e. they look like they will not bind together when turned.
6. Tighten the central bolt and then the two hexagonal bolts either side of the motor unit. You need to make sure that you do not over tighten the bolts. Finger pressure will suffice in most cases.
7. Switch on the Autostar and set the slew speed to a mid range setting 5 or 6. See Autostar section to find out how to do this.
8. Press the Up and Down arrows (or for RA Left and Right arrows) on the handset to operate the Dec or RA motor. Notice if the relevant axis responds quicker than before the motor alignment took place. Take special care to listen to the sound of the motor. If it sounds like it is making an odd noise, or it is straining to turn the axis then you must switch off immediately, or you will damage the motor. You will need to follow the above steps again and re-align the motor, ensuring the motor unit and the gears are less tightly coupled than last time.

Setting up the Telescope

9. Test the new alignment on either a terrestrial or astronomical object to determine the level of backlash still present if any. Use the backlash compensation utility on the Autostar to reduce the backlash even further (see earlier section in chapter).

Cog Adjustment of the Motors. Sometimes you might find that the RA or Dec axes do not turn at all, when you use any of the arrows keys on the Autostar Handset. This can be due to the cog not being firmly attached to the worm shaft inside the motor unit box. This is easily remedied by tightening the small grub screw on the side of the cog. Once the screw is tightened, the shaft will be able to turn the cog and hence, the motor attached to the shaft will turn the relevant axis.

Training the Motors (ECT: 10 Minutes)

In order to increase pointing accuracy of the Goto system the motor drives must be trained in both the RA and Dec axes. Training allows the Autostar to 'remember' inaccuracies in the motors and the mechanical gears and compensates for them. The procedure should be performed periodically (monthly) to maintain accuracy of Goto operations, whenever a different OTA is attached to the mount, or when problems occur.

The training utility is found from the following option:

| Setup→Telescope→Train Drive |

Useful tips follows which will aid you in carrying out the training of the motors:

- Remember to set the model number to your telescope in the Autostar handset, upon first initialization of the Autostar.
- Use a high power eyepiece such as a 12 mm reticle eyepiece with crosshairs. This will greatly improve centralization of the object.
- Change the slew speed to a slower setting when using the arrow keys, so that you don't accidentally shoot past the centre, when cantering an object. I generally use the Autostar slew modes 4, 5 or 6.
- To assist in training the drives, I produced a 'bulls-eye' drawing and placed it as far away as possible from the telescope (more than 100 m, the further away the better the accuracy). Used in conjunction with the crosshair eyepiece it can greatly enhance the accuracy of the training. You can create a 'bulls-eye' image (example shown in Figure 4.20) using most commercial software paint packages or an old fashion pencil and compass to draw concentric circles.
- Training can be done at night on a star. However, the only star that should be used for this procedure is the Pole star, Polaris as it moves very little. Using other stars will reduce the accuracy of the procedure as they move too quickly.
- Upon a full 'Reset' all training information is lost, so you will have to re-train again.

The procedure is as follows (to train the mount in Alt/Dec jump to step 8).

1. Press 'Enter' when the 'Train Drive' option is selected. The up/down scroll keys will alternate the display between 'Az/RA Train' and 'Alt/Dec Train' options.
2. First choose the 'Az/RA Train' option and press 'Enter'.

Figure 4.20. Bullseye drawing.

3. The Autostar will then display text asking you to point the telescope at an object at least 100 m away. This does not necessarily have to be an object on the horizon but it does have to be stationary. Centre the object in the telescope using the 26 mm eyepiece by moving the mount manually. This is done by unlocking the RA and Dec axes. Do not use the Autostar arrow keys.
4. Press 'Enter' when the object is visible in the eyepiece.
5. The Autostar will then display 'Centre reference object'. Use the arrow keys to centre the object in the eyepiece. Important note: Before you press 'Enter' you *must* move your eye away from the eyepiece. The reason is in step 6 below.
6. The whole telescope tube will suddenly start to slew for a few seconds and then stop. The object will have moved out of the field of view of the eyepiece. The Autostar will then display '*Press > until it is centered*'. Press the right arrow key > to centre the object in the eyepiece. (You must make sure that the object does not go past the centre of the eyepiece. If it does then the whole procedure will have to be repeated).
7. Pressing 'Enter' again will slew the telescope for a few seconds in the opposite direction to that which occurred in step 6. The Autostar will then display '*Press < until it is centered*'. Press the left arrow key < until the object is back in the centre of the eyepiece. The mount is now trained in Az/RA.

The next step allows you to train the mount in Alt/DEC. The process is similar to that of training the mount in Az/RA as described in earlier steps:

8. Use the scroll keys to display '*Alt/Dec Train*'. Press 'Enter' and repeat steps 3 to 6, this time using the up ∧ and down ∨ keys to centre the object each time the telescope slews away from it. When completed the mount will be trained in Alt/DEC.

Setting up the Telescope

Once you have trained the drives, if you want more accurate tracking of objects then you need to perform a Periodic Error Correction (PEC) of the motor drives.

PEC 'Smart-Drive' Training (ECT: 30 Minutes)

Although the motor and gearing system of the LXD is very accurate, it is not mechanically perfect. These imperfections can cause minor inconsistencies in the tracking of an object, where the object may appear to speed up or slow down seemingly at random, despite the motor gears running at a constant sidereal speed rate.

A PEC (or Smart Drive) training utility of the RA motor is carried out to teach the Autostar where the mechanical imperfections on the motor gears are located. PEC is carried out at night with a star.

Some points to note before carrying out the PEC training of the RA motor:

– Perform a calibration of the motors.
– Train the motors as outlined in the previous section of this chapter.
– Set up the telescope for Goto, ensuring the telescope RA motor is tracking.
– If you are observing from the northern hemisphere pick a star in the south region of the sky and at least 30° above the horizon to reduce atmospheric effects. Conversely pick a star in north region of the sky if you are observing from the southern hemisphere.

The following procedure describes how to perform PEC for the LXD motor:

1. Navigate to the Smart Drive menu option:

   Select Item: Setup→Telescope→Smart Drive

2. Remove any previous PEC data by selecting the 'PEC Erase' option and pressing 'Enter'.

   Select Item: Setup→Telescope→Smart Drive→PEC Erase

3. Press 'Enter' again to confirm the 'YES' option. The Autostar will then display 'Erasing...'. After a few moments the erasure is completed. This is indicated by the display reverting back to the 'PEC Erase' option.

4. Select the 'PEC Train' option under the 'Smart Drive Option'.

   Select Item: Setup→Telescope→Smart Drive→PEC Train

5. Press 'Enter' and the Autostar will display 'Center Star. Centre the star in the cross hairs of the reticle eyepiece and Press Enter'.

6. Press 'Enter' once again to start the procedure. The Autostar will display 'Training PEC'. A numerical count will be displayed. This represents the number of positions of the worm gear mechanism. After about ten minutes, when the count reaches 150, the cycle is completed and the Autostar reverts to the previous menu.

7. Peer into the eyepiece and use the Autostar arrow keys to keep the star in the centre of the field of view. It is best to use a 12 or 9 mm illuminated reticle eyepiece, to provide accurate star positioning. If you do not use a reticle eyepiece then you will

need to judge the position of the centre of the field of view and try to keep the star close to its centre. However, this is not an ideal solution, and PEC will not be as accurate as using an eyepiece with defined central markings. The speed should be set to 2x sidereal speed (key 2) to keep the star in the centre of the field of view. If a faster speed is required then you need to check the reasons why the star is drifting, so much. E.g. polar misalignment.

8. Once the training is completed you need to 'Park' the telescope by selecting the Autostar 'Park' option. The telescope's power will have to be switched off and then back on again in order for the new PEC data to take effect, and for Autostar to remember where it is on the worm gear.

PEC is a learning system and, so rather than fully retrain the Autostar every time with new data, the existing data can be refined more than once using the 'PEC Update' utility.

| Select Item: Setup→Telescope→Smart Drive→PEC Update |

Once Autostar has been trained with PEC the Smart Drive utility is activated as follows:

| Select Item: Setup→Telescope→Smart Drive→PEC On/Off |

When the PEC option is set to 'On', the Autostar finely adjusts the speed of the motors from its PEC training data as it encounters each imperfection of the gears. Hence, using PEC the tracking is more accurate. It is said to reduce the periodic error by as much as a third (from around 15 arc-seconds uncorrected to less than 7 arc-seconds with PEC). Like training data, all PEC training data is lost after a full Reset.

It has been noted by some LXD owners that PEC can introduce unnecessary drift in RA tracking into the LXD mount. Basically, experiment with PEC to see if the accuracy is truly enhanced. Probably the best way to provide precise tracking is through auto-guiding with a digital device.

Tripod Setup (ECT: 5 to 10 Minutes)

Setting up the tripod is the first thing do when setting up the telescope. This is a straight forward task which doesn't take very long. The most important thing that you need to remember is to make sure that the tripod is level with the ground. If the telescope is used on uneven ground and the tripod has not been levelled correctly, then you may find that the tracking or Goto feature will not be as accurate as it should be.

Tripod Ground Placement

Before you start placing the mount head and the OTA on the tripod, you need to make sure that the tripod is placed, such that the star emblem ☆ on one of the tripod legs is facing north (LXD55 tripod setup). This is done using either a magnetic compass

Setting up the Telescope

or an electronic compass device such as a GPS. You may have to use the magnetic compass some distance away from the tripod legs as ferrous (Iron) constructions could interfere with the direction reading. Further details about aligning the tripod are in Chapter 5.

Levelling the Tripod

The tripods for both LXD55 and LXD75 series are height adjustable. Each leg is individually raised or lowered in order to level the tripod. In fact all three do not necessarily have to be exactly at the same height to achieve a good levelling with the ground.

A small spirit level is very useful in helping you achieve an accurate level quickly and easily. For LXD55 tripods which come with the eyepiece tray, place the spirit level upon each of the three flat horizontal struts in turn.

Observe the bubble in the sprit level and adjust the height of each of the legs accordingly until the bubble floats towards the centre of the level ending up between two short line marks on the level. A good level is achieved when the spirit level is placed anywhere on the eyepiece tray and the spirit bubble is in the centre of the two level indicators.

For the LXD75 which has a built-in eyepiece tray as part of the central bolt, the spirit level is placed along any of the three horizontal bars in turn (Figure 4.21) and the legs adjusted in the same way as for the LXD55 tripod legs. The only difference for the LXD75 is that the rounded legs are locked in place with two bolts rather than a single bolt for the LXD55 tripod legs. You must ensure that the bolts which lock the tripod in place are sufficiently tight to avoid the legs accidental retracting when the rest of the telescope components are added to it.

It shouldn't take more than a few minutes to set the tripod level with the ground. Unless you are doing accurate imaging or tracking, a rough height adjustment is all

Figure 4.21. Spirit level on LXD75 eyepiece tray.

that's necessary to carry out simple observations. However if you are using the tripod for serious observing, then you will need to be more thorough with the levelling.

Once levelling is achieved, some people place marks on each of the tripods legs or, leave the tripod legs locked in their setup positions when the telescope is packed away. This is so that tripod can be set up much quicker next time. However, this is only a good way to set up the tripod, if you are going to be using the telescope in the exact same spot each time. For other locations you will need to level the tripod.

Finderscope Alignment (ECT: 15 to 30 Minutes)

Finding objects in the sky is a two stage process. First, you sight an object in the finderscope, and then look through the eyepiece of the main telescope to see if the object is there. Even with Goto systems, you will still need to use a finderscope to sight objects. This is normally done when the alignment setup procedure for the Goto function is performed at the beginning of an observing session. Telescopes have a small field of view and without a finderscope, you will struggle sighting those alignment stars in the eyepiece.

The LXD55 and LXD75 telescopes are supplied with an 8×50 finderscope. Single stranded cross hairs are viewed in the finderscopes' field of view. An object sighted in the finderscope is focused by twisting the finderscopes' end section. Some users tend to find that there can be a significant amount of resistance when attempting to twist the end section (LXD55 rather than the LXD75 finderscopes). The resistance varies from user to user, where some users (including myself) struggled to twist the end section, whilst others didn't have any problem at all. The resistance can be reduced by removing the finderscope tube assembly from its rings and repeatedly twisting the front section until the twist action is smoother. In actual fact, it will not be necessary to re-focus the finderscope, until maintenance is preformed on it.

Most finderscopes are held in place by a set of two slip rings. Three adjustable thumbscrews are located at equal distances around each of the rings 120° from one another. There are six thumbscrews in all. The positions of the thumbscrews are such, that they provide varying directions of movements, to precisely align the finderscope with the main OTA. Tightening or loosening the thumbscrews, causes the finderscope to point in slightly various directions. The direction of the object in the finderscopes' field of view is not always intuitive to the way the thumbscrews are adjusted. It will take time before you get the hang of the objects movement in the finderscopes' field of view, in relation to how the thumbscrews are tightened or loosened. It is quite common for users to completely undo one of the thumbscrews from one of the slip rings, whilst trying the alignment procedure. The key is to always make sure that there is plenty of room for adjustment for each of the thumbscrews, before alignment is carried out. This is done by making sure that there is enough thread protruding out on the inside of the ring to hold the finderscope, so that it is roughly in the centre of both the slip rings (Figure 4.22).

Alignment of the finderscope can be done either in daylight or during an observing session in the dark. I suggest that approximate alignment is done during daylight hours and fine adjustment is done on a star during an observing session. For daylight

Setting up the Telescope

Figure 4.22.
Finderscope and slip rings.

alignment, you will need to find an object that is at least half a mile away, in order to make the alignment as accurate as possible hence, requiring little adjustment when you take the telescope out at night to complete the alignment procedure.

Some points to note before you carry out the finderscope alignment procedure:

- Autostar is not needed unless you wish to use the slew function to move the two axes, in order to find a suitable object. Unlocking the axes and moving the telescope manually is sufficient.
- Use a relatively low power eyepiece for the main telescope. The 26 mm supplied is best used. Moreover, use an eyepiece that you will be using frequently.
- Make sure the finderscopes' base is firmly fixed to the OTA.
  1. Point the telescope to a distant object, at least half a mile away. Try to use an object that contains an outstanding feature, such as a cross on top of a church steeple or the tip of an aerial. Lock the axes when the object is in the field of view of the eyepiece in the main telescope. Since the field of view is small you will have to make fine movements of the two axes until you have the object near the centre of the field of view.
  2. Look through the finderscope and note the position of the object in relation to the centre of the field of view, where the cross-hairs intersect one another. Due to the low power of the finderscope, the object should still be in the same field of view as that of the eyepiece in the main telescope.
  3. Adjust the thumbscrews first on the slip ring closest to the front of the finderscope, to bring the object as close as possible to the central intersection of the cross hairs. You may have to resort to adjusting the other set of thumbscrews on the slip ring closest to the finderscopes' eyepiece end, into order to bring the object into the field of view. Most of the time, you will only need to adjust one set of thumbscrews on one of the slip rings, to get the object dead centre.

Alignment is successful when an object viewed at the intersection of the cross hairs in the finderscope, is also seen in the eyepiece of the main telescope (Figure 4.23). Different finderscopes can be used instead of the standards one supplied with the telescope. These are described later in Chapter 12.

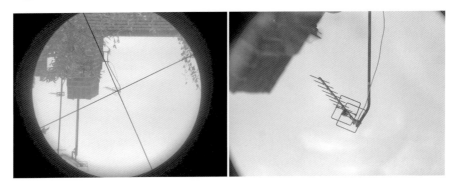

Figure 4.23. 8 × 50 mm finderscope FOV (left) and, AR-6 FOV with a 40 mm eyepiece (right).

Focuser Adjustments (ECT: 10 to 15 Minutes)

Most of the LXD series come with the same focusing mount, with the exception of the SCT. Checking the focuser to see if it operates correctly takes only a few moments.

1. Gently rack the focusing tube all the way out and then all the way in. Note how smoothly the mechanism performs.
2. Tighten or loosen the friction screw on top of the focuser until you sense that there is just the right amount of friction.
3. Now, use the standard eyepiece supplied in the focuser and once again gently rack the focus tube in and out checking if the movement is smooth.

If at any point whilst racking the focusing tube you encounter stiff resistance or jerky movements, check the grooves underneath the tube for dirt or faults in the grooves. There are four screws holding the shaft and cog that racks the focusing tube in and out. Undoing these screws, cleaning and re-greasing the cog usually resolves the problem (Figure 4.24).

Figure 4.24. Cog mechanism of focuser.

Setting up the Telescope

The focuser for the SCT is designed differently to that of the other telescopes in the LXD series, where the mirror is moved up or down, rather than using a focusing tube. There is no friction screw to adjust, as the focusing mechanism is internal to the OTA. Hence, the only check to carry out is to ensure that there is no significant shifting of the image in the field of view of an eyepiece, whilst adjusting the focusing knobs at the back of the SCT.

Conclusion

Tinkering with your telescope is part of the enjoyment of astronomy. Often astronomers try and rush assembling the telescope, especially if they have a limited time to observe due to the possibility of cloud cover and they want to slot in a quick observing session. Assembly should be carefully thought through and most importantly it shouldn't be rushed. In Chapter 11 I will discuss how to maintain your telescope so that it operates at its best each time you use it.

You should become familiar with assembling and dismantling the telescope indoors during the day before doing it outside in the dark at night. In the end setting up the telescope will become second nature leaving you to concentrate more on the observing aspects of astronomy.

CHAPTER FIVE

Polar Alignment and Goto Setup

One of the most important parts of setting up a telescope for a night's observing is polar alignment. All equatorial mounts with tracking abilities must have the RA shaft aligned with the celestial pole in order to accurately track the stars in the night sky. The greater the precision of the polar alignment, the more accurate is the tracking. Polar alignment is also a necessity for accurate Goto of celestial objects and to reduce rotation of field effects. Given the significance of polar aligning and setting up the Goto facility I have dedicated a whole chapter to it. You may find that methods for polar aligning a telescope described in this chapter may seem daunting to you at first and alignment may not be successful the first time round. However, with a little persistence and patience you will master the methods in no time.

Using Polaris to Find the North Celestial Pole

The nearest star to the north celestial pole (NCP) is Polaris in the constellation Ursa Minor (also known as the Little Dipper or Little Bear). The closeness of this star to the NCP, just under 1° away (1° is about twice the apparent diameter of the Moon), makes polar aligning very easy for observers in the Northern hemisphere. Figure 5.1 shows a time lapse photo of the North Celestial Polar region of the sky. The small bright point near the centre of the image is Polaris.

Polaris is probably the most looked at star in the sky; I say 'looked at' rather than 'observed' as most astronomers tend to only use it as a guide in finding the celestial

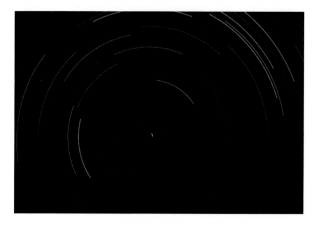

Figure 5.1. Star trails near north celestial pole. Image courtesy of George Tarsoudis.

pole rather than spending a few moments to study it detail. You can find Polaris by using two stars from the stars of the Big Dipper (the Plough) known as the 'pointers' shown in Figure 5.2. Follow an imaginary line from the 'pointer' stars Merak and Dubhe in the Big Dipper, and you will locate Polaris.

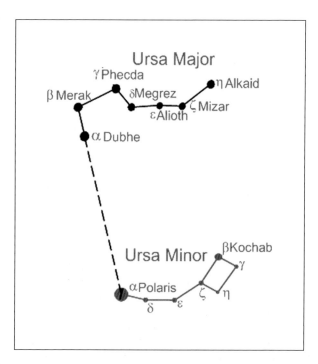

Figure 5.2. Finder chart for polaris. Image courtesy of Alan Marriott.

Polar Alignment and Goto Setup

Determining Latitude

I illustrated in Chapter 2 that the celestial poles are the projected points of the geographic North and South poles of the Earth. So, if you stood at the North Pole you will see Polaris virtually overhead, and if you were near the equator you would see the star very low down near the horizon. The projected height at which Polaris is above the horizon is your observing latitude.

You will need to determine the latitude of your observing location if you wish to polar align your telescope. The manual supplied with the telescope displays a list of latitudes of the major cities of the world. The values given are generally okay for setting the latitude of the telescope for polar aligning. There are other means of getting the exact latitude of your observing location, as the Autostar handset normally requires a more precise value.

Maps: Traditional printed maps provide lat/long information.

Computer programs: There are many Astronomy computer programs available, which give the precise latitude of many cities and places around the globe.

Internet: Map programs are available through the Internet. Entering the zip code provides a map of the local area as well as the latitude and longitude of the location. These values are usually listed with other information below the map. More powerful online map programs such as Google Earth and MSN Maps allows you to zoom in to your location and provides lat/long coordinates. See website list in Appendix D.

GPS: If you are fortunate to own a Global Positioning System (GPS) device you can use it to determine the latitude and longitude of your observing position. This means you will be able to roughly align the telescope mount at your observing location during the daytime, and fine adjust the alignment of the mount when Polaris is visible in the sky, later on in the evening.

Aligning the RA Shaft with the North Celestial Pole

Most telescopes nowadays that are mounted equatorially include a polar alignment telescope, which are built in to the RA shaft of the mount. The LXD55 and LXD75 series is no exception. The etched pattern in the viewfinder also illuminates with a red light, so that it can be seen in the dark.

Setting up the telescope so that the RA shaft is due north is a straightforward task. You will need a standard compass or GPS device to assist in obtaining an accurate North, especially if you are setting up the telescope during the day. The Earth's magnetic North and South poles are in different locations from that of the true geographic North and South poles so the compass will not provide a true accurate reading. Magnetic variation depends upon your location on the Earth and information (Libraries, Internet etc...) is freely available, which can help you determine the angular difference that the compass is pointing from true North. When taking a compass reading, you need to make sure that you are not in close vicinity of anything that might affect the reading, such as electric generators, power lines (overhead or buried), lumps of iron, ferrous material etc...

GPS is far more accurate as it does rely on the Earth's Magnetic field to provide a compass reading. For night-time preparations however, there are other methods to obtain polar alignment without resorting to a compass. I will talk about these other methods later in this chapter.

Before you carry out the alignment try to make sure that the tripod is level with the ground. Make sure you switch off the illuminated reticle of the polar viewfinder when you have finished the alignment task, otherwise you will drain the batteries. Also, it is likely that you will be performing the alignment in the dark, so make sure you have a torch handy.

Tripod Setup

LXD55 – All Models

Locate the star ☆ symbol on one of the legs of the tripod stand (Figure 5.3, left image). Place the tripod on the ground so that the star symbol faces due North. The best way to do this is to place a compass or GPS handset on the plastic tray in parallel with the leg that has the star symbol. Move the whole tripod around, until the compass reads due north or the GPS compass reading displays 000°N. For compass readings, make sure you take into account the local magnetic field variations and compensate accordingly.

LXD75 – All Models

There is no star symbol for the LXD75 tripod, so you should ensure that the leg which should face due North is the one which is directly beneath the protruding metal tip on the tripod stand (Figure 5.3, right image). When the Mount is attached to the top of the tripod you should notice that one side of the latitude adjusters and the fine control azimuth knobs are directly above the leg. The LXD75 stainless steel tripod will interfere with cardinal readings on the compass, if one is held too close to them. So, you will need to determine due north by keeping the compass away from the tripod. A GPS device is useful in these cases as the readings are not interfered with the metallic tripod.

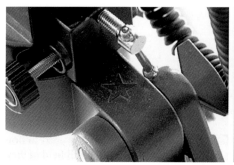

Figure 5.3. LXD55 (left) and LXD75 (right) tripod north pointers.

Polar Alignment and Goto Setup

Figure 5.4. LXD mount latitude scale.

Latitude Adjustments

Once you have set up the tripod, you can either attach the mount or the entire telescope assembly for polar alignment. It is best not to attach the OTA until after you have polar aligned the RA shaft. This is just in case you need to physically shift the entire telescope around, to get a suitable position for precise polar alignment. Once you have aligned the RA shaft and have attached the OTA to the mount, you should always make a quick alignment check to ensure that the RA shaft is still pointing to the celestial pole. The weight of the OTA will exert pressure on the latitude adjuster bolts, which in turn might shift the RA shaft marginally out of alignment.

The next step is to adjust the latitude of the mount to match the latitude of your observing location. Figure 5.4 shows the latitude scale on the side of the LXD mount.

The scale does not provide latitude values smaller than 2° increments, so you will have to approximate. Adjusting the mount is done using the push-pull latitude adjusters. As you tighten one end of the latitude adjuster and undo the opposite one, the RA shaft will raise or lower in angular height about a central pivot. The final latitude value is determined with an arrow symbol located directly above the scale. Once you have set the latitude you must tighten both sets of latitude adjusters to lock the mount in place. Note: this is a rough latitude setup. The latitude will be precisely adjusted when Polar alignment is carried out.

Locating Polaris in the Polar Viewfinder

The tripod stand and RA shaft should be set due North. The OTA should be rotated in Dec to the horizontal position to allow the polar viewfinder clear sight of the sky. You should be able to see Polaris in the viewfinder.

You will notice that Polaris is somewhat dimmer than expected compared to how you would see it with the naked eye. This is due to the viewfinder's small field of view and focal length. In light polluted skies Polaris will be the only star that stands out in the viewfinder whilst in dark skies there are surrounding stars that can be seen as well.

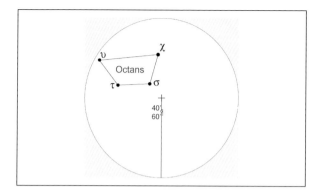

Figure 5.5. View through polar viewfinder (annotated).

The quadrangle pattern left of the central cross represents the stars of the constellation **Octans** in the Southern hemisphere. The little circle on the vertical line between the 40′ and 60′ markers represents the position of the Polaris in the Northern hemisphere, with respect to the NCP (Figure 5.5).

Adjusting the mount so that Polaris is in the little circle is sufficient for basic tracking and Goto operations for visual use. Without determining the exact orientation of Polaris with respect to the NCP, you may find inaccuracies or drift in right ascension and declination, whilst tracking the stars and it will not be good enough to carry out tasks such as CCD imaging or astrophotography. More stringent alignment methods are required.

In the course of 24 hours, Polaris scribes a small anticlockwise circle around the NCP, hence it is in different positions at different times. Therefore, you need to determine the position of Polaris in relation to the NCP at any given time if you want to obtain precise polar alignment.

One method is to use the RA Setting Circle and the silver calendar circle located directly beneath the RA setting circle (Figure 5.6). I will discuss the use of the manual RA and Dec axes setting circles in Chapter 7.

The calendar circle is set in 1 month gradated intervals 1 through 12, representing January, February, etc... The intermediate lines represent 10 day intervals and the smallest gradations 2 day intervals.

You will also notice that there is a gradated scale with East and West markings (E 20 10 0 10 20 W) located under the '12' and '1' monthly gradations values. This scale represents the 'Meridian Offset' and is used to set the longitudinal difference between your observing location and the nearest time meridian line. The lines are gradated in 5° intervals.

First, you need to know the longitude of your observing location to work out the offset of the meridian line. For example if you are in Washington DC (longitude 77° W) and the Eastern Standard/Daylight Time Zone reference meridian line is 75° W then you need to set the meridian offset by rotating the silver dial to 2° W to line up with the marker on the polar housing collar (between the 0° and 5° markers to the right of the 0 value). The meridian offset value will of course change if you travel to a location of different longitude.

Polar Alignment and Goto Setup

Figure 5.6. RA setting circle and calendar dial.

Now that you have set the calendar dial, the following procedure is carried out to determine the correct position of Polaris in relation to the NCP. *Note*: This procedure is also appropriate for Southern Hemisphere polar alignment.

1. Rotate the RA axis until the white line marker is lined up with the embossed arrow on the RA shaft. Lock the RA axis when done.
2. Unlock the RA setting circle lock knob and rotate the RA setting circle so that the '0' value is aligned with the white marker and the embossed indicator arrow (Figure 5.6). Lock the setting circle with the RA Setting circle lock knob. Do not rotate the RA axis for this step, only the setting circle. *Note:* If the meridian offset has been set then the '0' value of the RA setting circle will align up with the meridian offset value and not the '0' value on the silver calendar dial.
3. Unlock the RA axis and rotate it until the readings on the RA setting circle (current time) and the readings on the silver calendar dial (date) line up together. An example is shown in Figure 5.7. You must use UTC time and not daylight saving time (DST) to correctly set the RA values.
4. Use the adjusters of the latitude and/or azimuth controls to precisely place Polaris inside the little circle on the gradated line.

Another method to determine the orientation of Polaris with respect to the NCP is to use a star chart, planisphere or astronomical computer packages. All computer planetarium packages provide a zoom facility, which allow the user to obtain close up views of any area of the sky. Most of these planetarium packages are similar in many ways and so the following steps should generally be applied in order to provide the appropriate display of the celestial pole region.

1. Make sure that RA and Dec grid lines are set so that they are visible on the sky map.
2. Set the observing location and the time for when you plan to use the telescope. Make sure you have considered daylight saving time in your time setting.
3. Zoom in to the north polar region of the sky so that Polaris is displayed as well as the NCP which is where all the RA grid lines converge.
4. Print out the display for use at the time of polar alignment.

Figure 5.7. RA setting circle and calendar dial set to 4th October at a time of 23:50 local time (arrowed).

To align the mount, all you need to do is orientate the vertical line in the polar viewfinder by rotating the RA shaft, until the little circle on the gradated line matches the real position of Polaris with the NCP, as seen on the map.

Traditional Methods for Polar Aligning

Perfect polar alignment implies that the RA and Dec axes of an equatorially mounted telescope are exactly in parallel with the RA and Dec lines of the celestial sphere. Traditional methods such as the 'Two Star' method and the 'Drift' method are used by astronomers to polar align mounts, that do not have built-in polar alignment scopes. These methods are also suitable for polar alignment in the Southern hemisphere.

Please note, when you carry out either method described, make sure that the tripod is level and that you have placed Polaris in the Polar viewfinder to obtain a rough polar alignment. However, if Polaris is not accessible from your location i.e. hidden behind a tree or house, then use a compass or constellation to determine where Polaris should be. A rough estimation will be sufficient to start with.

Two Star Method

Direct alignment methods assume that Polaris is visible in the sky from the observing location. Astronomers tend to set up their equipment to face South so that they can observe objects in that area rather than observe in the North which means objects found facing North such as Polaris maybe hidden from view behind buildings or trees. Hence, alternate methods are used to polar align the telescope without the need to use Polaris at all.

This method bears no relation to the two star alignment method carried out by the Autostar, which will be discussed later in this chapter. It is a wholly manual method for polar aligning. The Two Star method uses stars lying at the same RA (or Dec) but at a fixed distance from one another. This, in effect aligns the RA and Dec axes

Polar Alignment and Goto Setup

Table 5.1. Two Star Method Alignment Stars

Season	Star Name or ID	Mag	RA	Dec
Dec Alignment (Stars with similar RA)				
Northern Hemisphere				
Spring	Star 1: Regulus (Alpha Leo)	+1.4	10h 08m 22s	+11° 58' 02"
	Star 2: Alpha Sex	+4.5	10h 07m 56s	−00° 22' 18"
Summer	Star 1: Pi Her	+3.2	17h 15m 02s	+36° 48' 33"
	Star 2: Delta Her	+3.1	17h 15m 02s	+24° 50' 21"
Autumn	Star 1: Scheat (Beta Peg)	+2.4	23h 03m 46s	+28° 04' 58"
	Star 2: Markab (Alpha Peg)	+2.5	23h 04m 46s	+15° 12' 19"
Winter	Star 1: Meissa (Lambda Ori)	+3.4	05h 35m 08s	+09° 56' 02"
	Star 2: Alnilam (Epsilon Ori)	+1.7	05h 35m 41s	−01° 12' 19"
Southern Hemisphere				
Spring	Star 1: Iota Aqr	+4.3	22h 06m 26s	−13° 52' 11"
	Star 2: Sadalmelik (Alpha Aqr)	+3.0	22h 05m 47s	−00° 19' 11"
Summer	Star 1: Alnilam (Epsilon Ori)	+1.7	05h 35m 41s	−01° 12' 19"
	Star 2: Meissa (Lambda Ori)	+3.4	05h 35m 08s	+09° 56' 02"
Autumn	Star 1: Alpha Sex	+4.5	10h 07m 56s	−00° 22' 18"
	Star 2: Regulus (Alpha Leo)	+1.4	10h 08m 22s	+11° 58' 02"
Winter	Star 1: Iota1 Sco	+3.0	17h 47m 35s	−40° 07' 37"
	Star 2: X Sgr	+4.6	17h 47m 23s	−27° 49' 51"
RA Alignment (Stars with similar Dec)				
Northern Hemisphere				
Spring	Star 1: Zaniah (Eta Vir)	+3.9	12h 19m 54s	−00° 40' 00"
	Star 2: Zeta Vir	+3.4	13h 34m 41s	−00° 35' 46"
Summer	Star 1: Deneb (Alpha Cyg)	+1.3	20h 41m 26s	+45° 16' 49"
	Star 2: Delta Cyg	+2.9	19h 44m 58s	+45° 07' 51"
Autumn	Star 1: Algenib (Gam Peg)	+2.8	00h 13m 14s	+15° 11' 01"
	Star 2: Markab (Alpha Peg)	+2.5	23h 04m 46s	+15° 12' 19"
Winter	Star 1: Aldebaran (Alpha Tau)	+0.9	04h 35m 55s	+16° 30' 33"
	Star 2: Alhena (Gam Gem)	+1.9	06h 37m 43s	+16° 23' 57"
Southern Hemisphere				
Spring	Star1: Nu Hya	+3.1	10h 49m 37s	−16° 11' 37"
	Star2: Eta Crv	+4.3	12h 32m 04s	−16° 11' 46"
Summer	Star 1: Lambda Vel	+2.2	09h 07m 59s	−43° 25' 57"
	Star 2: Sigma Pup	+3.2	07h 29m 14s	−43° 18' 05"
Autumn	Star 1: Zeta Vir	+3.4	13h 34m 41s	−00° 35' 46"
	Star 2: Zaniah (Eta Vir)	+3.9	12h 19m 54s	−00° 40' 00"
Winter	Star 1: Kaus Media (Delta Sgr)	+2.7	18h 20m 59s	−29° 49' 42"
	Star 2: Ascella (Zeta Sgr)	+2.6	19h 02m 36s	−29° 52' 49"

of the telescope with the RA and Dec axes of the celestial sphere. The greater the distance between the two stars the more accurate is the alignment. The Dec alignment determines whether the altitude of the RA shaft is at the same height as the celestial pole. The RA alignment procedure determines whether the RA shaft is shifted in azimuth to the left or right of the celestial pole.

There are very few bright stars in the sky which lie on the same RA or Dec. Therefore I have compiled a list of example stars in Table 5.1 along with their celestial co-ordinates for each season. Obviously the stars will be observed at high or low altitudes depending upon your observing latitude. You should note that both the RA and Dec co-ordinates for the two stars are not exactly the same, but their alignment will be satisfactory for observing and short exposure astrophotography.

All of stars in the table are found in star atlases, so you should familiarise yourself with the area in which the stars are located before you attempt to try and find them in the night sky. Try not to pick stars that are less than 30° or so from the horizon. Stars low down do not show their true position in the sky due to atmospheric refraction making them higher than they actually are.

Before you start, you should have Polaris in the field of view of the polar viewfinder. Use a high power eyepiece such as a 12 mm or greater; the greater the power used, the more accurate is the alignment. A 12 mm eyepiece with an illuminated reticle can be used to centre the star accurately in the field of view. If you don't have a reticle eyepiece use your best judgment to determine the central field of view of the eyepiece. Once you have aligned both stars switch to a higher power eyepiece such as a 6 mm.

Try to look through the eyepiece so that the telescope OTA is oriented such that the focuser's knobs are directly below you. This will provide an accurately determination of the location of alignment stars relative to the centre of the field of view. Achieving this orientation however, may put you in an uncomfortable or awkward position with your body if the star is high in altitude, so you will have to use your best judgment.

You should note that this is entirely a manual method therefore you should not adjust the position of the stars in the field of view using the Autostar handset. You should only use the latitude adjusters or the azimuth control knobs to centralize the star in the eyepiece's field of view.

RA Alignment (ECT: 10 Minutes). This is done with two stars with the same Dec but different RA.

1. Go to the first star, preferably the one with the higher declination.
2. Lock the RA axis, leaving the Dec to rotate freely about its axes.
3. Move the telescope in Dec ONLY to the second star and note its position relative to the central field of view.
4. If the position of the second star is to the *LEFT* of the central field of view then the RA shaft is too far west (left) of the NCP. Use the azimuth control knobs to laterally shift the whole telescope mount to the *RIGHT*, until the star is near the central field of view in the eyepiece.
5. Alternately, if the position of the star is to the *RIGHT* of the central field of view then the RA shaft is pointing too far east (right) of the NCP. Use the azimuth control knobs to laterally shift the whole telescope mount to the *LEFT*, until the star is near to the centre of the field of view in the eyepiece.

Dec Alignment (ECT: 10 Minutes). This is done with two stars of the same RA but with different Dec.

1. Find the first alignment star by manually moving the telescope about both axes.
2. Centre the star in the field of view.
3. Lock the Dec axes only, leaving the RA axes to rotate freely about its axes.

Polar Alignment and Goto Setup

4. Move the telescope in RA only until you come to where the second alignment star should be located and assuming the star is visible in the eyepiece note the relative position of the star to the field of view centre.
5. If the star is *BELOW* the central field of view then the altitude of the RA shaft is too far north of the NCP and needs to be *LOWERED* until the star is near the central field of view.
6. Alternately, if the star is *ABOVE* the central field of view then the altitude of the RA Shaft is too far south NCP and needs to be *RAISED* until the star is near the central field of view.

The Two Star method is time dependant, so you need to do it reasonably quickly (10 to 15 minutes) to avoid errors caused by field rotation during the exercise.

For both RA and Dec alignment methods, move the telescope back to the first star to check that it is still near the central field of view and repeat the procedure again, until both stars are reasonably near the central field of view when you alternate between them. This means that the RA shaft is aligned with the NCP. Table 5.2 provides a summary of the actions to perform for the two star alignment method.

The stars you pick will determine the accuracy of the alignment. In fact, if you find more than two stars with similar RA or Dec positions then by all means use them to improve the accuracy. This method is normally used before the more accurate Drift method is carried out. Once you have aligned the RA and Dec axes of the telescope with the real RA and Dec of the celestial sphere, then there should be little or no drift in the object's position in the field of view of the eyepiece and the telescope should be able to track objects for reasonably long periods.

Drift Method (ECT: 1 Hour)

The second method for polar alignment is by observing in the field of view the way a star drifts, whilst it is being tracked by the telescope. This is normally done using a reticle eyepiece with crosshairs, so that the relative drift is easily seen. The aim of the method is to reduce the drift of the guide star, by repeatedly adjusting the azimuth and the altitude of the mount, until eventually there is little or no drift for at least five minutes. This is the minimum amount of drift time to achieve accurate tracking for more than a couple of hours or so.

Guide stars are used, one on the local meridian (your due South) and one either due east or due west and around 15° to 20° above the horizon. These locations provide the greatest drift effect. The reticle eyepiece should be oriented, such that the crosshairs are

Table 5.2. Summary of Two Star Alignment Method

Axes	Star Position in Field of View	Action
RA		
	Star 2 is to the LEFT of centre	Shift azimuth mount to the RIGHT
	Star 2 is to the RIGHT of centre	Shift azimuth mount to the LEFT
DEC		
	Star 2 is BELOW central FOV	LOWER altitude of RA Shaft
	Star 2 is ABOVE central FOV	RAISE altitude of RA Shaft

Table 5.3. Drift Method Action Summary

Star Location	Star Drift Direction	Action
Meridian	Star drifts upwards	Move star to the right (Azimuth)
	Star drift downwards	Move star to the left (Azimuth)
East	Star drifts upwards	Move star down (Altitude)
	Star drifts downwards	Move star up (Altitude)
West	Star drifts upwards	Move star up (Altitude)
	Star drifts downwards	Move star down (Altitude)

aligned to the east-west and north-south axes of the mount and the guide stars should be positioned along the set of horizontal crosshairs. The east-west crosshair can be aligned, by switching the motor drive off and noting the drift of the star. Adjustments to the mount depend upon which way the star drifts in the field of view. If you see any drift, repeat the exercise and use the Autostar handset to bring the guide star back in line with the crosshairs.

You should note that the drift in the West guide star will require altitude adjustments in reverse to that of the East guide star altitude adjustments. Also, if you are using the LXD Newtonian for polar alignment then you will need to perform the actions in reverse to that specified in Table 5.3.

Setting up the Goto Facility

All telescopes with Goto have to be setup before they can be used to find objects in the Sky. Before any type of Goto operation is carried out you need to run through the Goto alignment procedure. The Autostar handset provides several methods to set up a telescope for Goto operations.

- Easy
- One Star
- Two Star
- Three Star

The procedure for setting up the Goto facility using the Autostar handset is universal across the entire LXD range. You will need to use the finderscope to assist you with the Goto setup, so you need to make sure that you have it aligned with the main telescope. Also, the motor drives should have been trained prior to Goto setup otherwise the alignment might fail.

I would highly recommend that you carry out a 'dummy' test run during daylight hours, just to monitor how the telescope reacts to the various alignment methods stated in the following sections. Of course, testing in daylight means you have to trust that the telescope is pointing in the right direction to the star. You have to blindly

Polar Alignment and Goto Setup

follow the instructions on the Autostar's display screen, until you get an alignment message successful. If you consistently get alignment failure messages, then you will need to carry out particular checks, such as the motors. See Chapter 13 for possible causes and solutions of alignment failure.

Easy Align (ECT: 5 Minutes)

Select Item: Setup → Easy

The Easy Align option is designed to make alignment a simple step process. By simply following the instructions on the handset, two stars are automatically picked by the Autostar and the telescope then slews to them. Sounds simple enough! However, the Autostar is notorious for picking stars that are obstructed behind buildings or trees. In these cases when the Autostar suggests a star, press the scroll down key on the Autostar and it will pick another star to align on.

A disadvantage of using the Easy Align method is that the Autostar tries to pick stars whose proper names are assumed to be known to the user. If you are not familiar with every proper name in the Autostar database (this includes myself!), then you will struggle to know which star the telescope is supposed to point towards. So, to help you remember the names of the stars used by the Autostar and where they are in the sky I have provided a list of stars in appendix A along with simple finder charts. If you want more detailed charts, then I would recommend using a computer Planetarium program to display the proper names on star maps on the screen.

One Star Alignment (ECT: 5 to 7 Minutes)

Select Item: Setup → One Star

Although the method is called one star alignment, there are actually two stars involved; one is Polaris and the other star is chosen by Autostar, hence the method is very similar to Easy alignment.

The one star method is mainly suitable for those who want a quick set up in order to use the Goto facility, but are not fussed if tracking an object is not perfect. Hence, the one star method is really only suitable for visual observation.

Unlike other methods, the one star method requires a physical adjustment of the mount in order to get Polaris in the centre of the field of view. When the telescope is in the Polar Home Position it is actually quite difficult to get Polaris dead centre in the field of view of the eyepiece. If the RA shaft is perfectly polar aligned, Polaris tends to move around the edge of the field of view when the telescope is rotated around the RA shaft, with the Dec axis fixed at $+90°$. Therefore, adjusting the altitude of the mount in order to get Polaris in the centre of view, the RA shaft will be offset by a small amount from the NCP, so tracking of objects will not be as accurate and the object will drift out of view after a period of time.

Two Star Alignment (ECT: 5 to 6 Minutes)

Select Item: Setup → Two Star

This is very similar in method to the easy alignment already discussed earlier in this chapter.

Which Stars to Choose

For the majority of the time, I tend to choose the two star alignment method over Easy alignment, as it gives me the freedom to choose which two stars I know will be visible from my location. I try to choose stars that are not hidden behind trees or buildings and whose names and sky positions I am familiar with.

You should try to pick two stars that are at least a quarter of the sky away from one another, but not on opposite sides of the sky. This is contrary to what the general consensus suggests. The reason is as follows. Misalignment of the RA axis to the North celestial pole produces an offset known as 'the cone error'. The cone error is where stars near to the North Polar region of the sky are more closely aligned to the RA axis, than stars on the opposite side of the sky. In fact, the cone error changes due to the sky rotating around the north celestial pole as the night progresses and Goto becomes less accurate. Hence, choosing stars that are in the same hemisphere of the sky will mean that the offset of the stars, due to Polar misalignment, will be more or less similar to each other, than choosing stars on the opposite sides of the sky. Autostar will be able to compensate for the offset and you will be able to use Goto accurately (for a couple of hours at least). You can reduce the cone error further by performing a three star alignment. The Autostar software includes a third star alignment utility which I will discuss next. To eliminate cone error you must have accurate Polar alignment which is discussed in Chapter 5. Another type of cone error occurs if the OTA and the RA axes are not perfectly parallel to one another. This can be rectified by carrying out the OTA alignment procedure outlined in Chapter 5.

It is normally good practice to pick alignment stars in the area of the sky that you will be observing. Finding objects using Goto in the area where the two stars are picked during the alignment setup, tends to be more accurate than finding objects outside that area on the opposite side of the sky.

Three Star Alignment (ECT: 6 to 7 Minutes)

Select Item: Setup → Three Star

This method is similar to Easy alignment where the Autostar automatically select the stars to be used for alignment, but an additional third star is selected which lies on the opposite side of the sky to the first two stars.

With the three star alignment method you should in theory, be able to slew to the opposite sides of the sky objects with greater accuracy, than if the other alignment

Polar Alignment and Goto Setup

methods were used. In practice however, you may find that unless you have set up the telescope precisely, the Goto accuracy degrades over a period of time due to cone error. Hence, you may have to perform a three star alignment every two to three hours in order to maintain Goto accuracy during the night.

Alignment Setup – Success or Failure?

Once you have chosen your two stars, now the fun part begins. I say 'fun' without being too cynical, but successfully setting up the Goto facility using the two star alignment method can sometimes be a bit of a 'hit' or 'miss' affair.

Around 90% of the time you will get that all important 'Alignment Successful' message appearing on the Autostar display. It is followed by a 'beep' sound, signifying that the Goto setup is complete. However sometimes, for a multitude of reasons, the 'Alignment Failure' message appears on the Autostar display instead. Unless you know exactly the reason why the alignment failed, trying again is basically a fifty-fifty chance that it may pass or fail. Alignment is about ten percent action and ninety percent perseverance, if it all works okay then off you go and observe, but if it all goes wrong then you could be in for one of those nights where nothing seems to go right. There are many possible reasons for alignment failure and some are discussed in Chapter 13.

Suggestions below can be applied to any of the Goto alignment methods mentioned in this chapter.

1. It is best to use the 26 mm eyepiece that's supplied with the telescope as it provides the best magnification suitable for accurate Goto setup. A higher magnification (hence smaller field of view) means that it could take much longer to find the star. On the other hand, a wide-field low power eyepiece could make the Goto less accurate, if you subsequently switch to a higher power eyepiece.
2. For the first star that is picked, there is a possibility that the telescope might point to region of the sky which could be far from the actual position of the star. The misalignment could be due to various factors such as motor problems, incorrect location, date setting or others, which are outlined in Chapter 13. Slew the telescope until the star is in the centre of the field of view.
3. Autostar, during the procedure sets the slew rate to maximum which is the default setting. This is fine if the star is quite a distance away from where the telescope is pointing. However, once you get the star into the field of view of the eyepiece, it can be difficult to fine-adjust the position of the star to centre it. The slew rate during the procedure should be set to 5 or 6 (64× to 128× sidereal rate) in order to move the star small distances in the eyepiece. You can set the slew rate in the normal manner, without causing interruption to the alignment process – press the desired number on the keypad (see Chapter 6 for more details about Slew Speeds).
4. For the second alignment star that is chosen, if the telescope's position is a distance from the actual position of the star (just as with the first star alignment) then there

is a distinct possibility that the alignment will fail. Failure in this case depends upon how far the telescope is from the first and second star's actual positions.

There are cases where alignment is successful, even though in both the first and second star alignment the telescope is offset by about the same distance through a consistent error. From my own personal experience of Goto set up I have witnessed the telescope point to a region of the sky several degrees from star's actual position and the alignment was still successful!

Southern Hemisphere Alignment

Aligning the telescope RA to the South Celestial Pole (SCP) is a more difficult task to achieve than for the North Celestial Pole because there is a lack of significant bright stars close to the SCP. The nearest star to the SCP is Sigma Octantis which lies about 1.5° away. It is magnitude +5.5 which makes it barely visible to the naked eye, quite faint in comparison to Polaris.

Just like setting up the telescope in the northern hemisphere you will need to determine your observing latitude which will also be the projected height of the SCP. You will also need to level the tripod and set the RA shaft so that it points due South. Both the latitude and the direction can be found in the same way for finding the latitude and direction for the northern hemisphere. When the latitude and the direction are set this will be the Polar Home Position for the southern hemisphere.

The view through the polar viewfinder shows a quadrangle pattern etched in the field of view. This pattern represents four stars from the constellation of Octans, notably Sigma, Tau, Chi and Upsilon Octantis (Shown in Figure 5.5). To locate the SCP you will need to orientate the RA shaft, such that the quadrangle pattern in the viewfinder matches that of the four stars in the sky. When the stars are matched against the pattern, the cross at the centre of the field of view will represent the SCP. The quadrangle pattern will be in a different position around the SCP at different times of the year, so you will need to determine where the pattern will be when you carry out polar alignment.

Alternative methods for aligning the RA axes with the SCP include the Drift method and traditional Two Star method, which were discussed earlier in this chapter.

The Autostar will automatically determine whether you are in the Northern or Southern hemisphere from your location settings. The Goto alignment methods on the Autostar are the same as the Northern Hemisphere with the exception of the one-star alignment method as it uses Sigma Octantis as one of the alignment stars instead of Polaris. This is done by adjusting the calendar and RA dials on the RA Setting circle as described earlier in this chapter.

Non-Goto Operation of the Telescope

You don't have to have the Goto set up if all you want to do is track stars. As long as you have performed a polar alignment you can use the telescope for non-Goto operations,

Polar Alignment and Goto Setup

just like a traditional telescope with tracking features. There are two ways to do this; you can either unlock both RA and Dec axes and push the telescope to the desired position, using the finderscope to find to locate the object, or you can use the motors to slew the telescope to the desired position using the Autostar handset and the arrow keys.

The LXD mount has no slow motion controls unlike other non-Goto GEM mounts so once you have located the object you will have to use the Autostar arrow keys to slowly move around the area you are observing. Finding objects without the use of Goto is discussed in Chapter 7.

To set up the telescope for non-Goto operations perform the following steps.

Note: For firmware versions v4.n or later there is a new set of initial menu options which allows you to either perform an Automatic alignment by pressing '0' or skip the alignment by pressing the 'Mode' key.

1. Power on the Autostar
2. Press the 'Mode' key. Skip the date and time and DST settings.
3. Skip the Alignment setup. Press the 'Mode' Key upon seeing the 'Easy Align' display.
4. Goto the Targets option. **Select Item: Setup → Targets**
5. Select the 'Astronomical' Option to 'On'.
6. The RA motor will engage and the telescope starts tracking.

Conclusion

The most common fault the user finds with their telescope is when Goto alignment procedures are carried out. The telescope might have problems getting the Goto to work correctly, or it might locate objects perfectly every time. It is a common misconception that accuracy of Goto is directly linked to the tracking of stars. Accurate Polar alignment is only really essential if you wish to carry out astrophotography or CCD imaging, where precise tracking is necessary.

The main appeal of the telescope is the Goto facility and if that feature is no longer accessible then, you have to resort to manually pointing the telescope to find objects in the sky, which is not the reason why you purchased this particular telescope in the first place.

CHAPTER SIX

First Night's Observing

The night when you take your new telescope outside for the very first time can be a memorable experience to say the least, especially if you are a beginner. It could leave you either very happy with your purchase, or bitterly disappointed wondering whether you made the right decision to buy the telescope in the first place. First light of a telescope determines whether the optics are good enough to observe objects in the night sky, as well as determining whether the Goto facility is operating correctly. You'd rather hope that your first night will be successful, so you should try to be as organized as you can. Be prepared however, that things could go wrong the first time.

An Experience to Forget...

You've read the instruction manual from cover to cover and think that you are confident enough to set up the telescope in the dark. However things don't turn out the way they should be.

It looks like a clear night so you decide to take you new telescope outside and set it up for the first time. You've picked your location and are ready to set up the telescope. You seem to spend ages trying to level the tripod but no matter how many times you try to adjust the height of the legs, you can't seem to get a good level. In the end, you decide that the level you have achieved is good enough for first light observing. The mount is then attached. Whoops! Watch out for that counterweight! You attempt to polar align the RA shaft to Polaris. Well, you think its Polaris! it seems to be dimmer than normal when seen through the Polar viewfinder. You attach the tube assembly almost using will alone to hold the scope up in the air, whilst you blindly fumble with one hand to tighten the screws to fix the cradle assembly to the top of the mount.

Thinking that the worse bit is over you decided to plug in the Autostar and power it on (hoping that the batteries will last the session!). You set the time (is your watch correct?), date and location settings and prepare for Goto setup. You decide to go for the easiest setup option; 'Easy' align but by sheer bad luck the first star Autostar happen to pick is hidden behind the only tree in the garden. Selecting the next star from the Autostar list provides a star name that you are not familiar with. You decide to forego 'Easy' alignment and opt for the two star alignment method instead. You scroll through the list of named stars until you come across one that you recognize and is visible in the sky at the time. Upon selecting the star, the telescope begins to slew to the area which the star is located, but for some reason the telescope has stopped some distance short of the star. You have to slew it manually, until you centre the star in the field of view of your eyepiece. Picking your second star, the telescope slews across to the chosen area. However, once again the telescope appears to be pointing some distance away from the star. You manually centre it in the eyepiece again and press 'Enter' on the Autostar. The Autostar displays the sentence 'Calculating...'. It is the moment of truth! Have you done enough to successfully setup the telescope for Goto? The answer is a few moments later. 'Alignment Failure...'. What did you do to warrant such a failure? After cursing for a few moments you decide to try one more time... alignment failure again. You decide that you'll forego the Goto setup for another time and attempt to observe some objects in the night sky whilst it's still clear, by moving the telescope manually. After several minutes you manage to locate Mars and start to take one long look at it through the eyepiece. But as luck would have it, the clouds roll in for the evening...

Even experienced astronomers have had one of those nights when things simply do not go right. The best thing to do about it is to go right out and do it all over again. Practice makes perfect!

Location! Location! Where to Observe

It is very important that you try and survey the area where you are going to set the telescope up. You should try to find a location that will allow you to set the tripod up as level as possible with the ground. The air above concrete tends to be more turbulent than grass, so you should try to avoid patios and garden paths. Grass however tends to become damp, so it is best to try and keep most of the equipment away from the ground as much as possible.

If you are going to be observing from a remote location, try to visit there during daylight hours. Make a note of steps, ditches and any obstructions that you might trip over or fall into, when you are out there in the dark. In Chapter 2, I described what you need to prepare when you go out to observe.

Finally, if you are observing in a public place, make sure that you have permission to observe there, and if possible, notify the appropriate authorities in advance.

First Night's Observing

Figure 6.1. Case containing essential observing equipment.

Equipment! What to Take Out

You should try to plan ahead before an observing session so you know what equipment to take out. It's so easy to consider the 'just in case' factor and take too much equipment out; especially if it takes a while to set it all up. Be realistic and only take what you really want for an observing session.

When you are familiarizing yourself with the telescope for the first time you will only need a basic set of equipment to take out:

- The telescope (of course!)
- 26 mm eyepiece supplied with the telescope
- Higher power eyepiece such as 9 or 12 mm
- Star charts

As you become more confident with the operation of the telescope you may want to take out more equipment. I used to take out two large metal cases, containing various bits and pieces which on many occasions ended up using only one or two items from each of the cases. Nowadays, I have rationalized all my equipment and currently use a small metal case, which contains eyepieces I use the most frequently along with other essential accessories, such as the Autostar handset and a set of filters (Figure 6.1).

Any user of a telescope would agree that they often have to 'tinker' with their telescope to maintain its peak operating performance. The tools are necessary 'just in case' the telescope doesn't operate the way its intended to. A hardware toolbox can be used to contain tools and accessories. It also has handy storage compartments ideal for small items such as Allen keys and batteries (Figure 6.2).

Of course if you wish to carry out CCD or Webcam imaging, then you will need to take out with you all necessary equipment that goes with this type of activity such as

A User's Guide to the Meade LXD55 and LXD75 Telescopes

Figure 6.2. Large tool box.

laptops etc. In the end you could have so much equipment it could take you half the night to set it all up! I would strongly recommend that you practice setting up all the essential equipment during daylight hours to cut down the time it takes to set it up when you observe.

Telescope Setup

All LXD telescopes follow the same procedure for setup. It should take around 10 to 20 minutes to fully set up the telescope from assembly to Autostar Goto operation. Ideally you should have practiced the setup during the day, so you have some confidence in carrying it out at night in the dark.

1. Telescope assembly: Choose a suitable location and assemble the telescope (Chapter 4).
2. Polar align: Align the RA shaft with the celestial pole (Chapter 5).
3. Plug in Autostar: Connect the coiled lead from the Autostar handset to the HBX socket on the control box of the LXD mount.
4. Add power: Connect power lead from the power source (battery or mains). Switch on the power.
5. Remove start messages: When you power on the Autostar it displays 'Getting Started' and 'Sun Warning' messages. To stop these messages displaying at startup:

> Select Item: Utilities → Display Options → Getting Started → Off

> Select Item: Utilities → Display Options → Sun Warning → Off

You can change the scroll speed of the messages by pressing the up or down cursor keys. I tend not to have the messages displayed upon Autostar power up, as it means extra keys to press when setting up the telescope for observing.

First Night's Observing

6. Setting location, date and time: It is important that you try and get the time as precise as possible as it plays an important part of the Goto setup. Precise location is not as important and you can get away with the closest city to your location specified in the Autostar location database. If you are however a very long way from the city specified in the Autostar, then you need to provide a more accurate location of your position (discussed in Chapter 5).

    Of course, you can set up your own custom locations and populate Autostar with the precise latitude and longitude co-ordinates. There are several ways of determining the precise latitude and longitude of your location. These are described in Chapter 5. Autostar prompts for the location only when it is started up for the very first time. However if you have already operated the handset during the day then you should have set up your observing sites beforehand.

    Using a GPS add-on expedites the setup procedure. The GPS automatically inputs the precise location, date and time into the Autostar.

    The following procedure may vary slightly for different GPS devices.

    a. Plug the GPS add-on into the RS232 socket of the Autostar.
    b. Power on the Autostar.
    c. Skip the date and time initialization Autostar menu options.
    d. When the alignment setup option is displayed, press the button on the GPS receiver unit. A few moments will pass whilst the receiver locks onto a GPS satellite, and uplinks the relevant date, time and location into the Autostar.
    e. Unplug the GPS receiver and continue with the Goto alignment process.

    Hence, if problems occur with the Goto alignment you can eliminate the possibility that the problems are caused by incorrect Time, Date or Location information.

7. Goto setup: The next step is to choose your method of alignment for Goto setup (Easy, Two Star or Three Star). For the first night-time use of the telescope you should try 'Easy' alignment. If not, try Two Star alignment. You may need to refer to the finder charts in Appendix A to assist you with the alignment.

8. Alignment success! Goto an object: Once the Goto alignment setup is successful, you should be ready to slew to an object. Try to choose an object within the area of the two stars that you or the Autostar selected during the alignment procedure. The reason is that it is more likely the telescope will locate the object precisely in the field of view. The accuracy is greater between the two stars than elsewhere in the sky. Three star alignment improves GOTO accuracy for the whole sky (see Chapter 5).

Slewing the Telescope

Autostar has the ability to change the speed of the motors. There are nine speeds available where each speed is accessed through the numerical keys on the Autostar keypad. Table 6.1 shows the speeds.

When a numerical key is pressed the slew speed is momentarily shown on the Autostar, before the display changes to the current action.

Table 6.1. Slew Speed Rates Available on Autostar

Slew Mode Selection	Sidereal Rate	Arc-min/sec	Deg/sec
Key 1	1x (Tracking Rate)	0.25	0.004
Key 2	2x	0.5	0.008
Key 3	8x	2	0.033
Key 4	16x	4	0.067
Key 5	64x	16	0.27
Key 6	128x	30	0.5
Key 7	384x	90	1.5
Key 8	768x	180	3.0
Key 9	1152x	270	4.5

The fastest sew speed is notorious for being very noisy when in operation. Although, this speed means that the telescope reaches its target very quickly, it can put a strain on the motors, reducing their operational life. At 3 a.m. in the morning, the telescope slewing at maximum speed sounds like a hammer drill being used, much to the annoyance of the neighbours and their canine companions alike!

Fortunately, Meade introduced the 'Quiet Slew' option into Autostar firmware updates v2.6Ea onwards which put paid to those noisy Goto slews in the middle of the night. The 'Quiet Slew' option sets the maximum slew rate to 1.5 Deg/sec which is equivalent to Key Number 7 in the slew speed list.

The 'Quiet Slew' option is found through the Setup menu:

> Select Item: Setup → Telescope → Quiet Slew

Goto alignment procedures with the 'Quiet Slew' option always set to 'On' will slew the telescope at the slower speed than if the maximum speed was used. The telescope however will take longer to reach its target at the slower speed but, it's a compromise to extend the operational lifetime of the motors as well as getting some relative peace and quiet during observing sessions.

Slew speeds vary depending upon the task you are carrying out. For example, centering an object in the field of view of the eyepiece requires fine movements implying the lowest slew speeds such as 1 or 2 slew mode selections, whereas manual adjustment of the telescope's position requires 7 or 8 slew mode selection. Table 6.2 lists suggested slew rates.

Table 6.2. Slew Modes Selection for Various Tasks

Task	Slew Mode Selection
Centering object in Finderscope	6–7
Centre Alignment Star during Goto Setup	5–6
Goto Setup	Automatic Highest Slew Rate 8–9
Object Fine Adjust	2–3
Object Guiding (for Imaging)	3–4
Terrestrial Viewing	5–6 (3–4 for fine centering in FOV)
Training the motors	4–6

First Night's Observing

After a while you will become familiar with which slews speed to use for the appropriate task.

End of the Night – Packing Up

Packing up the telescope at the end of an observing session is normally performed much quicker, than when you first set up. However the task should still be planned out with some thought in mind. Some astronomers speedily 'throw' equipment into cases/containers and sort out where each individual component goes later, in the comfort and warmth of their own home. Others meticulously place each component in their respective positions, which of course takes longer. Somewhere in between there is a happy medium for packing equipment away in a relatively short amount of time, without damaging anything or yourself in the process.

There is no hard and fast rule for packing away your LXD telescope and associated equipment. Suffice to say that I tend to start off with the smaller equipment and work my way to the bigger and heavier components. I also tend to put away first the most expensive equipment, such as laptops and CCD cameras, although it does depend upon where you pack up, such as in a private garden or public place.

Dismantling the telescope should not be a rushed task. Keep in mind that you are dealing with a bulky, heavy and expensive piece of equipment, so handle each component carefully as you pack it away.

Packing Up in Public Locations

Public places of course provide less security than a private location and sometimes it is beneficial to have a second person to look after the equipment whilst you pack it all away. Astronomers tend to be a trustworthy bunch. Especially where there are large gatherings such as star parties. But if lots of public attend, then it's easy for an eyepiece to go missing in the dark. So be vigilant and be aware of the equipment that is left on view.

Parking the Telescope

If your telescope is permanently housed or you are not intending on moving it overnight, then there is an Autostar feature that assists you when you are finished for the night. Parking sets the telescope back to the Polar Home Position before you power off the handset. The Park option is found via the Autostar Utility menu:

> Select Item: Utilities → Park Scope

When the Park Scope option is selected, the telescope will immediately start slewing to the Polar Home Position. This is regardless of the position of the telescope when the

Park option was first selected. The slew speed is determined from whether the 'Quiet' slew mode is enabled or not. When the Polar Home Position has been reached the Autostar then requests that the power is switched off. A point to remember is that you cannot stop the parking procedure once it has started, so the only course of action is to power off the Autostar. This is why you must make sure you have finished observing for the night, before the parking is carried out.

Parking also provides a true test that the Polar Home Position has been set up correctly, (the Dec axis should end up pointing to 90° when the park action has been completed).

Try to remember to step back from the telescope and keep out of its way when it starts to move. This happens immediately after the 'Park Scope' option has been selected and the 'Enter' key is pressed. You should also make sure that cables do not get in the way whilst the telescope is in motion.

Parking the telescope is useful if you want a fast setup. This is because the original alignment settings are saved from the previous observing session, so repeating the alignment procedure is not necessary. All you need to do is power on the handset, provide the Date and Time information and go straight to an object. Simple!

Sleep Scope

Select Item: Utilities → Sleep Scope

The Sleep Scope function is useful for when you want to keep the telescope switched on but leave it in a dormant state until you want to use it again. This could happen if you want a break from observing or if clouds appear and are likely to clear later.

Upon activation, Sleep Scope switches off the tracking facility and dims the illumination on both the Autostar screen display and key pad, like a screen saver. All training and alignment settings as well as the date and time internal clock settings are retained. The Autostar is essentially set in a standby power-save mode.

To reactivate the telescope, all you need to do is press any key on the Autostar keypad. The tracking will automatically start up again. You should note that the 'Enter' key should not be pressed to reactivate the Autostar if the last menu option displayed upon reactivation happens to be the 'Sleep' option. Pressing 'Enter' will activate the 'Sleep' function again and put the telescope back into a dormant state. If your LXD is permanently housed, it is tempting to use the 'Sleep' function to keep the telescope switched on so you don't have to switch off the Autostar. However I would advise against it and you should use the Parking utility instead.

A final note on the Sleep Scope utility; upon reactivation from the standby state especially if it has been several hours in standby, the last object that was used for Goto will still be designated in the Autostar memory. Pressing the Goto key will slew the telescope to that object. If the object happens to be below the horizon, Autostar will report it so.

Summary

Expectations always run high on the first night of using a telescope, especially if there is a lot of new technology for you to play with. You tend to want to try everything out all at once to see what the telescope can do.

So was your first night a successful one? The consensus from the majority of LXD telescope owners I spoke to was a resounding 'yes'. The unlucky few owners who had one of those nights when it didn't go to plan, suggested that things did improve in subsequent observing sessions. In either case, it was certainly a learning experience they would not forget.

CHAPTER SEVEN

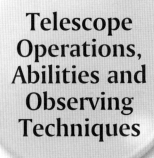

Telescope Operations, Abilities and Observing Techniques

Observing in Comfort

When you observe with an equatorially mounted telescope such as the LXD, you may find that you operate it differently to an Altazimuth mounted telescope such as a Meade LX200 or LX90. This is especially evident when it comes to obtaining a comfortable viewing position.

For example, if you observe an object with an LX200 SCT near the zenith and then another object near the horizon, you will notice, assuming you have set up the tripod correctly, that the eyepiece is more or less kept at head height. However, with an LXD SNT or the LXD Reflector, the focus mount is located near the top of the tube assembly. Therefore, you may discover that it is difficult to reach the eyepiece comfortably for objects almost directly overhead, especially if you are height challenged.

Conversely, observing objects near the zenith with long focal length refracting telescopes (such as the LXD AR-5 or AR-6), where the focuser is located at the bottom end of the OTA, may result in you crouching down almost to the ground, in order get into a position to be able to look through the eyepiece.

You may need to adjust the tripod to a suitable height to achieve a comfortable viewing position. In addition, use a small stool or bench to stand on in order to reach the eyepiece when the telescope is pointing to a position high in the sky. The Autostar has a useful utility called Max Elevation. This allows you to set the altitude limit that the OTA can slew to, so that you don't have to stand in an awkward position to look at an object overhead. You can manually slew past that limit if you so wish. The utility is also used for LXD refractors to prevent the focusing end of the OTA from accidentally colliding with the tripod mount, causing motor damage if the OTA is pointing almost straight up. The only disadvantage of using the Max Elevation utility is that you will not be able to see any objects above the elevation limit, without manually slewing past that limit.

Select Item → Telescope → Max Elevation

Observing Celestial Objects Across the Meridian

The Meridian is an imaginary arc that travels from due north to due south passing through the celestial pole. For an Altazimuth mounted telescope simple up-down movements will locate objects on either side of the Meridian, without putting the telescope into an unusual orientation. This is because of the way the altitude line is parallel to the Meridian line. So, the Altazimuth mount will smoothly follow it.

Pointing at objects on either side of the Meridian with an LXD equatorially mounted telescope is somewhat different. You will not find it simple to move the telescope across the meridian to reach its target compared with an Altazimuth mounted telescope. This is because the RA axis is physically fixed in the plane of the Meridian line. You will need to 'flip' the OTA from one side of the RA axis to the other side in order to obtain a suitable observing position (Figure 7.1). In fact, if an object is very close

Figure 7.1. Viewing the same star from opposite sides of the meridian.

Telescope Operations, Abilities and Observing Techniques

to the meridian, it can be observed with the telescope positioned in two different orientations.

When using the Autostar to slew to objects across the meridian, the software automatically orientates the telescope to what it thinks is the most suitable position for the observer to view the object. This means that if you have commanded the Autostar to find an object across the meridian, it will take longer than normal to slew to that object. The telescope has to re-orientate itself to its optimum viewing position for that area of the sky.

How Faint Can You See – Limiting Magnitude

When I tell people that I own a telescope, the response I normally get is 'How far can you see with it?' Well, the answer I normally provide is "millions of light years, and it depends upon what object I am looking at, e.g. a Star or Galaxy". However, I try to explain that it is not the case of how *far* you can see with a telescope but rather how *faint* you can see with it. Faintness is not necessarily a measure of the distance of an object. A faint object could be small or have a low surface brightness.

A telescope's visual performance is governed its physical characteristics such as aperture size and the quality of the optics. Other factors include sky conditions and even the observer's visual acuity to see minute details. Some of these factors, such as sky conditions will vary from night to night, and the telescope's visual performance will change over a similar period.

Although the Autostar contains thousands of objects in its database, many of the objects are too faint to see with an LXD telescope without electronic assistance such as CCD imaging.

The limiting visual magnitude of a telescope is the faintest theoretical magnitude that can be seen under ideal sky conditions. See Table 7.1.

The photographic limiting magnitude is different to the visual limiting magnitude. Cameras and CCD devices (see Chapter 10) have the ability to integrate (add) the light from images, thereby enhancing the light grasp of the telescope.

Table 7.1. Limiting Visual Magnitude for All LXD Models

Telescope Model	Visual Limiting Magnitude	Photographic Limiting Magnitude
6" N Reflector (152 mm)	13.5	16.0
5" AR Refractor (127 mm)	12.8	15.3
6" AR Refractor (152 mm)	13.5	16.0
6" SNT (152 mm)	13.5	16.0
8" SNT (203 mm)	14.0	16.5
10" SNT (254 mm)	14.5	17.0
8" SCT (203 mm)	14.0	16.5

Data courtesy Meade.

Table 7.2. Resolving Power of the LXD Telescopes

Telescope Model	Resolving Power (arc-sec)
6" N Reflector (152 mm)	0.74
5" AR Refractor (127 mm)	0.90
6" AR Refractor (152 mm)	0.74
6" SNT (152 mm)	0.74
8" SNT (203 mm)	0.56
10" SNT (254 mm)	0.45
8" SCT (203 mm)	0.56

Data courtesy Meade.

How Much Detail Can You See – Resolving Power

A telescope's resolving power (related to Dawes Limit or Rayleigh Criterion) is the theoretical limit for which it can split point-like sources such as stars. It is similar to 'Resolution', which is the ability of a telescope to discern fine detail on an object such as the Moon or a Planet.

A telescope's ability to resolve stars depends upon the diameter of the telescope's aperture (Table 7.2). The larger the aperture the more it can resolve detail. However, if you observe a binary or double star with a telescope whose resolving power is greater than their separation, then the stars cannot be split into their individual components, no matter how much you magnify the object.

In reality, atmospheric conditions make the theoretical limit difficult to achieve. The resolving power under typical sky conditions for the naked eye is about 1 arc sec. Even under ideal sky conditions telescopes such as the LXD55/75 10" SNT will not achieve sub 0.5 arc sec resolutions due to atmospheric turbulence.

How Far Can You Zoom – Magnification Power

Magnification is calculated by dividing the focal length of the telescope by the focal length of an eyepiece. For example, a 20 mm eyepiece on an LXD55/75 8 inch f/10 SCT (2032 mm) will produce a magnification of 102× (2032/20 mm).

There is a practical limit for the magnification of a telescope. A general rule of thumb is to multiply the factor 2.4 by the aperture of the telescope (in mm). This is the maximum useful magnification. For example, the most useful maximum magnification for an LXD55 6 inch (152 mm) refractor is about 360×.

Increasing the magnification of an object reduces the object's brightness (see also Exit Pupil) and lowers its overall contrast. Hence, high magnifications are normally confined to objects such as the Moon, Planets or Double Stars, rather than objects with low surface brightness such as Galaxies or Nebulae. Most practical observations

Telescope Operations, Abilities and Observing Techniques

with LXD telescopes are carried out using eyepieces whose magnifications are between 60× and 200×.

The Autostar contains a useful utility which calculates the magnification and field of view of an eyepiece from a selection it has in its list. The values displayed are dependent upon the telescope model and focal length, originally selected in the Autostar.

Select Item → Utilities → Eyepiece Calc. → Magnification

Select Item → Utilities → Eyepiece Calc. → Field of View

Scroll down the list of eyepieces and press 'Enter'. The Autostar displays the magnification or the field of view.

How Much Light from an Object Can You See – The Exit Pupil

The exit pupil of a telescope is the size of an image projected by the eyepiece onto the observer's pupil. It is calculated, by dividing the focal length of the eyepiece by the focal ratio of the telescope. Alternatively, the exit pupil is calculated, by dividing the diameter of the objective by the magnification used. So, for example, a 26 mm eyepiece with an LXD75 8 inch f/4 SNT (203 mm) will produce an exit pupil diameter of 6.5 mm.

Why is the exit pupil significant for observing objects though a telescope? Well, a person's pupil is naturally wider at night than during the day in order to gather as much light as possible. The pupil size for a relaxed eye during daylight hours is approximately 4 mm, whilst at night it dilates to about 6 mm (see Dark Adaptation). For direct visual observing, the exit pupil should be at least the same as the size of the eye's pupil in order to collect the greatest amount of light from the object.

For example, viewing an object through an LXD75 10 inch f/4 SNT with a 40 mm eyepiece, an exit pupil of 10 mm is produced. This is greater than the optimum night-time pupil size (6 mm). By using this eyepiece, over half the amount of light will not reach the eye's retina from the object, and so it will appear dim. However, a satisfactory bright image could still be produced, but only if the object is very bright initially. Experiment with different eyepieces until you are satisfied which ones to use, without compromising quality of the image and its brightness.

Observing Techniques

Dark Adaptation

To get the best out of observing your eyes need to be able to resolve details on objects under low light conditions. The time it takes for the eye to reach optimum sensitivity depends upon each individual, but on average it takes between twenty minutes to half an hour for the eyes to chemically adjust, so that they can be sensitive in the dark.

Sometimes astronomers find it difficult to get completely dark adapted. It could be that their observing location has too much surrounding light from a nearby residence, poorly directed street lighting, or a Full Moon present increased the background light levels of the sky. Some astronomers use a black cloth hood draped over their heads and the focuser, in order to mask out extraneous light from the surroundings. Other sources such as bright white light torches can quickly destroy several hours of dark adaptation. If red light is used, it should be of sufficient low intensity to not affect the light sensitivity of the eyes.

Physical factors affect seeing ability are age, level of sugar in the blood, diet and the relaxed state of the observer. It has been said that alcohol in small doses heightens visual acuity, but too much alcohol has the opposite effect (although it does instantly *double* the amount of objects you see during the observing session!). Of course, I would never recommend that you *drink* and *drive* (your telescope!) and common sense should prevail. Consuming copious amounts of Carrots is another misconception that is supposed to increase visual acuity in a person. There is no scientific evidence that it is true, unless the person already has a severe Vitamin A deficiency.

Some astronomers go to great lengths to speed up the dark adaptation process. Before carrying out observations, sunglasses are used to try and force the pupils to dilate, so that when they are taken off in the dark, the eyes have adjusted to the new lighting levels. I know several colleagues who have used this method with mixed success.

Keeping your eyes closed is not the best way to dark adapt either as you need to have your eyes open for them to focus on objects in low light conditions. The best way is to simply do nothing and let the eyes relax by themselves. Typically though, if you are setting up your telescope in the dark using just a red torch to assist you, then you will likely to be dark adapted, by the time you have completed the set up process. Hence, you can start observing straight away.

Averted Vision

When observing faint diffuse objects such as Galaxies or Nebulae, you may struggle sometimes to visibly see them in the field of view. This is because the eye is not as sensitive at the center of the retina than at its edge.

The retina contains two types of light detecting cells; cones and rods. The color sensitive cones are bunched together at the center of the retina where they are surrounded by rods, which are more sensitive to light. Under normal conditions, light from an image is projected directly onto the cones. By directing your eyesight just off to one side of the object, the image will be projected onto the light sensitive rods. It takes a little practice to view objects at the peripheral of your vision, as the eye will tend to try and focus back onto the object in its own center of view. You may see the object fleetingly in the corner of your eye as you look away from the object and back again.

Telescope Operations

This section deals with the various operational aspects of using the LXD telescope.

Tracking Objects

Not all celestial objects travel at the same rate across the sky. Solar system objects such as the Moon, Planets, Asteroids and Comets travel at different speeds. They will appear to move independently to the background of stars. Autostar compensates for these different speeds by altering the tracking speed of the motors. The tracking rate is found as follows:

Select Item: Setup → Telescope → Tracking Rate

There are three options available:

- Sidereal – default tracking rate
- Lunar – for tracking the Moon
- Custom

Sidereal Rate

The *Sidereal* tracking rate is the default speed used by the Autostar for tracking Stars (including the Sun), Galaxies and Nebulae.

Lunar Rate

The *Lunar* tracking rate is used if you are going to observe the Moon for more than a few minutes. Earlier versions of the Autostar firmware meant that you had to manually change the tracking rate, each time a Goto of the Moon was performed. However, latest versions of the firmware are smart enough to automatically change to the Lunar tracking rate when a Goto of the Moon is carried out. Upon selecting an object other than the Moon to Goto, Autostar will revert back to its previous tracking rate (Sidereal in most cases).

You should be aware however, that if you manually set the tracking rate to Lunar whilst the telescope is already tracking the Moon, the default rate from then on will be the Lunar tracking rate. The next object you Goto will track at the Lunar rate and not the Sidereal rate. In this case you will need to set the tracking rate back to Sidereal.

Custom Rate

The *Custom* rate setting allows you to enter a user defined value that matches the object's speed you are observing. This is as long as it is less than $2\times$ the Sidereal rate. The tracking rate can be advanced (+) or retarded (−) relative to the Sidereal rate:

Advancing the Tracking Rate.

1. Check that the current rate is set to Sidereal. This is done by pressing 'Enter' at the *Telescope: Tracking Rate* Option and notice if the '>' symbol is next to the *Sidereal* option.

2. Scroll through the options using the menu keys until the *Custom* option is displayed and press 'Enter'.
3. The Autostar will display '*Enter Rate Adj.*'. Enter a three digit value of your choice (see later in this section). Note that the symbol on the left of the screen is a plus sign '+' to represent that the value entered advances the Sidereal tracking rate.
4. Press Enter again to set the new tracking rate.

Retarding the Tracking Rate.

1. Set the tracking rate to *Lunar*. This is done by pressing 'Enter' at the *Telescope: Tracking Rate* menu item, scrolling down to the *Lunar* option and pressing 'Enter' again. The text *Telescope: Tracking Rate* is shown on the Autostar display.
2. Press 'Enter' again to see the tracking rate options and then scroll down using the menu keys until the *Custom* option is displayed.
3. Press 'Enter' and you will see 'Enter Rate Adj.' text on the top line of the screen. A pre-defined value of −35 is shown. This value is fixed for the Lunar rate.
4. Enter a three digit value of your choice (see later in this section). Note that the symbol on the left of the screen is a minus sign '−' to represent that the value entered retards the Sidereal tracking rate.
5. Press Enter again to set the new tracking rate.

There are almost 2000 rate settings from −999 to +999. Each unit value represents 0.1% of the Sidereal rate. So for example, if you set a value of +500, the tracking rate will be 50% faster (0.1 × 500) or 1.5 times the Sidereal rate. Understanding the negative values is less trivial. For example, a value of −300 represents a speed that is 30% slower than the Sidereal rate. To put it another way, it is 70% of the Sidereal rate. Table 7.3 highlights the custom rate range as a relative rate of the order, between 0.01 and 1.99 of the Sidereal rate.

To reset the tracking rate either set the tracking rate option to *Sidereal*, or enter 000 as the rate value in the *Custom* option.

Table 7.3. Sample Values of the Custom Rate Option

Custom Rate Value	Percentage (+advance / −retard)	Multiple of Sidereal Rate
+999	+99.9%	1.99 (199% Sidereal)
+500	+50%	1.5 (150% Sidereal)
+100	+10%	1.1 (110% Sidereal)
000	0%	1.0 (Exact Sidereal)
−035	−3.5%	0.965 (Average Lunar)
−300	−30%	0.7 (70% Sidereal)
−500	−50%	0.5 (50% Sidereal)
−900	−90%	0.1 (10% Sidereal)
−950	−95%	0.05 (5% Sidereal)
−999	−99.9%	0.01 (1% Sidereal)

Telescope Operations, Abilities and Observing Techniques

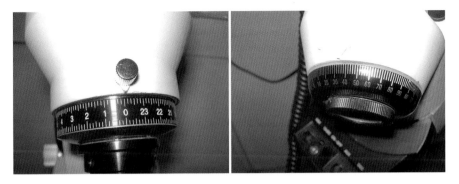

Figure 7.2. LXD75 mount setting circles. RA (left) and Dec (right).

Using the Telescope Without Goto to Find Objects

You may think that the non-Goto use of the LXD implies that it takes longer to find objects. This can be the case in some circumstances. However, astronomers have been using traditional means for finding objects in the sky long before Goto was introduced. The experienced astronomer will find objects in the sky just as quickly, using a sophisticated state-of-the-art computerized telescope.

Star hopping is a great way of learning the sky, whilst you attempt to find faint fuzzy blobs such as Galaxies or Nebulae. Start with a bright star that is easily locatable with your telescope. Then use sky maps to 'home in' to the object, 'hopping' from one star field to the next, until you locate the target. Star hopping is ideal for coming across objects on your way to locating the target. However, it normally takes longer to find the target, than simply pushing a few keys on the Autostar, and allowing the telescope to do all the work for you.

You can also use setting circles to find objects. Simply 'dial' up the object's RA and Dec coordinates to match the values of the setting circles on the telescope mount (Figure 7.2). The Dec setting circle should have been calibrated when the Polar Home Position was first set up. The RA values however change with Sidereal time, so you have to work out the RA range that is visible for any particular night of the year. Calibrating the RA setting circle is done as follows:

1. Find a bright star in the sky. Preferably one which is at least 30° above the horizon.
2. Look up the RA and Dec celestial co-ordinates of the star by either finding it in the Autostar database or from a star list.
3. Point the telescope to the bright star, either by unlocking the RA and Dec axes and moving the telescope manually, or by using the Autostar slew arrows. It is best to use a moderate power eyepiece such as a 15 or 12 mm and center the star in the field of view. The Dec co-ordinate should match that of the star's known Dec value.
4. Set the RA co-ordinate on the telescope mount first by loosening the small metal lock screw above the RA setting circle and then turning the RA setting circle to match

4. When the Goto button is pressed the telescope will not slew to the object directly but will first slew to a bright star nearby. The Autostar will display "*High Precision – Slewing . . .*".
5. When the star is reached, Autostar will display the named star and ask you to center it in the eyepiece. E.g. "*Ctr. Alnath. Press Enter*".
6. Upon pressing 'Enter', the telescope will start to slew to the Messier Object. A beep sound is heard to notify that the Goto action has been completed.

The HP mode once activated will *always* go to a bright star before the primary target is located. You may want to set the HP mode only when necessary, otherwise it will take longer to reach your chosen target.

Digital Setting Circles

Autostar can display the RA and Dec values of its current position. To display the RA and Dec co-ordinates hold down the Mode key for more than 2 seconds (Figure 7.4).

Digital setting circles are calibrated during the Goto alignment setup. The accuracy depends upon the precision of polar alignment and the success of the Goto alignment setup. A check to see if the setting circles are displaying reliable values, is to point the telescope to a star and compare the Autostar digital setting circle readings with that of the star's known RA and Dec co-ordinates. If you are confident that the Autostar is displaying accurate readings of RA and Dec co-ordinates, then you can use the digital setting circles to find objects. Use the arrow keys to find the object using its co-ordinates.

You can also enter the RA and Dec co-ordinates manually into the Autostar handset. This is done as follows:

1. Ensure the Alignment procedure has been carried out and that the telescope is tracking.
2. Press and hold the 'Mode' key for more than 2 seconds, release and use the scroll keys to display the current RA and Dec co-ordinates. (See Appendix C for list of information available in this mode).

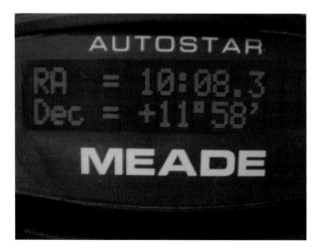

Figure 7.4. RA and Dec co-ordinates on Autostar.

Telescope Operations, Abilities and Observing Techniques

Figure 7.2. LXD75 mount setting circles. RA (left) and Dec (right).

Using the Telescope Without Goto to Find Objects

You may think that the non-Goto use of the LXD implies that it takes longer to find objects. This can be the case in some circumstances. However, astronomers have been using traditional means for finding objects in the sky long before Goto was introduced. The experienced astronomer will find objects in the sky just as quickly, using a sophisticated state-of-the-art computerized telescope.

Star hopping is a great way of learning the sky, whilst you attempt to find faint fuzzy blobs such as Galaxies or Nebulae. Start with a bright star that is easily locatable with your telescope. Then use sky maps to 'home in' to the object, 'hopping' from one star field to the next, until you locate the target. Star hopping is ideal for coming across objects on your way to locating the target. However, it normally takes longer to find the target, than simply pushing a few keys on the Autostar, and allowing the telescope to do all the work for you.

You can also use setting circles to find objects. Simply 'dial' up the object's RA and Dec coordinates to match the values of the setting circles on the telescope mount (Figure 7.2). The Dec setting circle should have been calibrated when the Polar Home Position was first set up. The RA values however change with Sidereal time, so you have to work out the RA range that is visible for any particular night of the year. Calibrating the RA setting circle is done as follows:

1. Find a bright star in the sky. Preferably one which is at least 30° above the horizon.
2. Look up the RA and Dec celestial co-ordinates of the star by either finding it in the Autostar database or from a star list.
3. Point the telescope to the bright star, either by unlocking the RA and Dec axes and moving the telescope manually, or by using the Autostar slew arrows. It is best to use a moderate power eyepiece such as a 15 or 12 mm and center the star in the field of view. The Dec co-ordinate should match that of the star's known Dec value.
4. Set the RA co-ordinate on the telescope mount first by loosening the small metal lock screw above the RA setting circle and then turning the RA setting circle to match

Figure 7.3. LXD55 RA fine adjustment scale.

the star's RA known value. The small arrow directly above the setting circle is used to point to the appropriate RA value. For LXD55 mounts, there is an additional RA scale to the left of the small arrow (Figure 7.3). This is used to finely adjust the RA in 1 minute intervals. The LXD75 mount does not have this additional scale.

5. Tighten the RA setting circle lock knob to lock it in place.

Test the setting circle configuration by pointing to a different star and comparing the RA and Dec reading on the telescope mount with the star's known Ra and Dec values.

The only negative aspect to using manual setting circles on the mount is that you have to recalibrate the RA circle at the start of every observing session. Digital setting circles can also be used via *Autostar*, which is discussed in the next section.

Using Goto to Find Objects

With Goto, objects in the sky are within easy reach with just a few key presses. Simply select an object in the database and press the Goto key. Of course this assumes that Goto alignment has been carried out perfectly every time. However, there will no doubt be times when the Autostar will struggle to find a faint object. This is where some very useful features on the Autostar are used to assist in finding those elusive objects.

- Spiral Searching
- High Precision
- Digital Setting Circles
- Synchronize

Telescope Operations, Abilities and Observing Techniques

Spiral Searching

Say for instance you want to find the star cluster M35 in Gemini. You select the object from navigating the Autostar menus:

1. Select Item: Object → Deep Sky → Messier Objects
2. Type 0 3 5 using the numerical keypad. Then press 'Enter'
3. 'Messier 35' is displayed on the screen. Press Goto. The 'Slewing...' message is displayed.

After the beep sound, which signifies that the Autostar has completed its Goto action, you might find that the familiar fuzzy image of the star cluster is not in the eyepiece. You have two courses of action to consider; the first is to slew the telescope using the Autostar arrow keys around the area, until you find the star cluster. Secondly, you perform a spiral search of the area. I personally find that spiral searching has more control over finding objects, than blindly slewing the telescope in random directions using the arrow keys.

A spiral search is carried out by holding down the Goto button for more than a few seconds. Looking at the field of view through the eyepiece, you will notice that the telescope will start moving around the area at a slow speed, sweeping the area in a square-like pattern. Once you have seen the object drift, pressing any key briefly will stop the spiral search. You can then center the object with the arrow keys.

The compactness of the spiral pattern can in fact be modified. This is not commonly known and I only discovered it one night by accident. I own two LXD telescopes one of 1200 mm (LXD55 AR-6) and one of 2000 mm (LXD75 SC-8). I accidentally failed to remember to change the telescope settings on my Autostar when I was using the SC-8 one night. The previous night I was using the AR-6. I noticed that the slew rate of the spiral search for the refractor was faster and had a wider coverage, than if I set the Autostar telescope setting to the SC-8. By 'tricking' the Autostar into thinking that a telescope of a different focal length is being used, the spiral search speed and the area coverage of the search can be modified. The higher the focal length the tighter and slower the slew of the search area and for wider and faster searches, focal lengths as low as 500 mm can be used. You must remember though, to set the focal length back to the original setting before you continue with your observing. You should also be aware that briefly pressing the Goto button for less than 2 seconds will start the Goto search action again.

High Precision

The High Precision (HP) feature provides a means for Autostar to obtain the precise location of an object. It does this by finding a bright star and synchronizing to it, thereby increasing the pointing accuracy around the area of the sky where the star is located.

1. Access the HP mode feature:
   Select Item: Setup → Telescope → High Precision
2. Set the HP Mode to 'ON'.
3. Select a target to find such as Messier 35 located in the constellation of Gemini.

4. When the Goto button is pressed the telescope will not slew to the object directly but will first slew to a bright star nearby. The Autostar will display "*High Precision – Slewing...*".
5. When the star is reached, Autostar will display the named star and ask you to center it in the eyepiece. E.g. "*Ctr. Alnath. Press Enter*".
6. Upon pressing 'Enter', the telescope will start to slew to the Messier Object. A beep sound is heard to notify that the Goto action has been completed.

The HP mode once activated will *always* go to a bright star before the primary target is located. You may want to set the HP mode only when necessary, otherwise it will take longer to reach your chosen target.

Digital Setting Circles

Autostar can display the RA and Dec values of its current position. To display the RA and Dec co-ordinates hold down the Mode key for more than 2 seconds (Figure 7.4).

Digital setting circles are calibrated during the Goto alignment setup. The accuracy depends upon the precision of polar alignment and the success of the Goto alignment setup. A check to see if the setting circles are displaying reliable values, is to point the telescope to a star and compare the Autostar digital setting circle readings with that of the star's known RA and Dec co-ordinates. If you are confident that the Autostar is displaying accurate readings of RA and Dec co-ordinates, then you can use the digital setting circles to find objects. Use the arrow keys to find the object using its co-ordinates.

You can also enter the RA and Dec co-ordinates manually into the Autostar handset. This is done as follows:

1. Ensure the Alignment procedure has been carried out and that the telescope is tracking.
2. Press and hold the 'Mode' key for more than 2 seconds, release and use the scroll keys to display the current RA and Dec co-ordinates. (See Appendix C for list of information available in this mode).

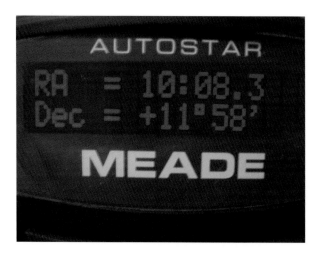

Figure 7.4. RA and Dec co-ordinates on Autostar.

Telescope Operations, Abilities and Observing Techniques

3. Press the Goto key, '*Object Position*' will be displayed on the top line of the Autostar screen.
4. Enter the RA values using the numerical keypad, Press 'Enter'.
5. Enter the Dec values then Press 'Enter'. The telescope should immediately start slewing to the designated co-ordinates. If this does not occur, then press the Goto key to start the slewing.

You can use the Arrow keys to go forward and back along the row at anytime if you wish to change the values.

Object Synchronization – The Sync Function

The synchronization or 'sync' function is used to compensate for alignment inaccuracies obtained during the Goto setup process. To synchronize an object:

1. Perform a Goto of an object and center it in the field of view of the eyepiece. Alternatively you can center any object without using the Goto facility.
2. Once you have centered the object in the eyepiece press the 'Enter' key and hold it for more than 2 seconds, then release it. You will see the message '*Enter to Sync*' displayed on the Autostar screen and a beep will sound.
3. Press 'Enter' again, another beep will sound and the screen will display '*Synchronized*'.

The 'Sync' function tells Autostar the exact location of the object without modifying any of the original settings, such as the RA and Dec co-ordinates. It is a means of increasing the pointing accuracy around the area of the object that is synchronized.

The 'Sync' function can also be used as an alternative to the one, two and three star alignment process.

The procedure is as follows:

1. Make sure that the telescope is in Polar Home Position.
2. Find a named star in the Autostar database and using the arrow keys slew to the star.
3. 'Sync' on the star.

You should synchronize on individual stars rather than objects such as the Moon or Planets, which move at slightly different rates to that of stars. Synchronizing is a quick and easy way to achieve Goto alignment. However, it is not intended to be used to enhance the pointing accuracy on objects located in other areas of the sky, so the area that you observe is limited around the object that you have "Sync'ed" on. Therefore it is good practice to 'Sync' on almost every object that you find.

CHAPTER EIGHT

The Universe at a Touch of a Button

Introduction

Observing astronomical objects in the sky is an awe-inspiring experience. The ever changing sky reveals myriads of objects within reach of an LXD telescope.

This book has so far covered the practical aspects of using the LXD telescope. If you are a beginner in using a Goto telescope, understanding its operation is your primary goal. Once you have picked up the basics, then the night sky is yours to enjoy.

Guided Tour

The Autostar Guided Tour is a useful utility. It allows the software to suggest which objects to observe. The utility is found by scrolling through the top level options.

Select Item: Guided Tour

The Guided Tour menu is split into the following options:

- Tonight's Best
- Stars of the Night
- How Far is Far

Selecting a tour takes you on a journey across the night sky. You will come across the many different types of objects that the Autostar has to offer. Each object encountered is accompanied with information to make the guided tour that much more enjoyable.

An object in the tour can be selected by using the up and down arrow menu keys. Guided tours can be customized and added to the menu list. The Meade website provides information on how to create a custom list and upload it to the handset (see Chapter 9 for more details).

If you are confident that the Autostar will accurately find any object, then I would certainly recommend that you try out one of the Guided Tours.

The Autostar Object Database

The Autostar extensive database contains thousands of objects to choose from like Stars, Planets and Galaxies, to more exotic objects such as Quasars and Black Holes. The database even lists stars that have planets around them. Of course some of the objects listed, such as Black Holes are difficult to observe even through the most powerful telescopes around the World (or in Space), let alone through a humble LXD telescope!

The Autostar Object menu is split into various libraries:

- Solar System
- Constellations
- Deep Sky
- Star

Other libraries are also available:

- Satellites
- Landmarks
- User Objects

Each menu option is further sub-divided into individual objects such as Planets and Named Stars as well as Messier and NGC deep sky catalogs. The next sections describe what you can see with your LXD telescope, with help from the Autostar database. Refer to Appendix B for extensive object lists such as named stars and deep sky catalogs.

The Solar System

Select Item: Object → Solar System

Before we talk about the rest of the objects in Solar System, let's discuss viewing the Sun through a telescope.

Observing the Sun Safely – Some Very Important Advice

Astronomers go to great lengths to ensure that their eyes are adequately protected when observing the Sun. Whether you are using the Solar Projection method or Solar filters, you should always take extra special precautions whilst you are observing the Sun. Please heed the public service announcement Sun warning.

The Universe at a Touch of a Button

> *Public Service Announcement:*
> Warning: *NEVER* look at the Sun through a telescope without the correct type of protective specialized solar filter as it will result in permanent eye damage and blindness.

This warning is probably the most important piece of advice any astronomer will ever give you, so please take the warning very seriously. Even, the Autostar handset displays a '*Sun Warning*' message, when it is powered on for the first time from out of the box. If you have any doubt in your mind that the equipment you are using is not sufficient enough for carrying out solar observing safely. *Simply don't do it* and seek proper advice.

The Sun – Solar Observing

Our closest star is a fascinating object to observe. Before dedicated filters were commercially available the only way to see the Sun's surface was through *Solar Projection*. However, with professionally manufactured solar filters now freely available, astronomers literally see the Sun in a different 'light'.

Solar Projection. The Solar Projection method is still commonly practiced, as it is especially useful for demonstrating the appearance of the Sun to large groups of observers at once.

The method is very simple; with the telescope OTA pointing towards the Sun, the solar disc is projected onto a piece of white card, which is placed at a fixed distance from the eyepiece (Figure 8.1). Try to use a simple un-cemented eyepiece, as the concentrated heat focused by the optics may otherwise cause damage to it.

Figure 8.1. Solar projection through an AR-6 refractor.

The aperture of the OTA is usually 'stopped' down to two to three inches using a home-made mask, in order to improve the contrast and reduce the intensity of the image.

The best telescope to use for solar projection is a Refractor for the following reasons. Firstly a Refractor is sealed from the front, so it will minimize the air currents within the tube, which is the main cause for reducing the clarity of the solar image. Secondly, since the light from the Sun passes straight through the OTA there are very few internal surfaces to reflect off, hence the build up of heat inside the OTA is kept to a minimum. Telescopes with multiple internal reflecting surfaces will suffer from heat damage caused by the Sun's intense heat. Therefore, you should *never* use an SCT or SNT for direct solar projection. A solar filter place over the front of the OTA must be used for these telescopes instead.

Solar Filters. Most astronomers nowadays use Solar Filters to directly observe the Sun through a telescope. They are usually made of glass and coated with special metallic material, designed to reduce the light and heat (infra-red) intensity from the Sun by more than 99.9%, before it enters the telescope. These filters allow astronomers to view the Sun in 'white light'.

Hydrogen Alpha filters are commercially available. They reveal features on the Sun not normally seen with white light solar filters, such as solar prominences. However, these filters are much more expensive. In this section I will discuss what you can see through a white light solar filter rather than a Hydrogen Alpha filter.

Mylar sheets are available such as Baader Astro Solar Film. Such a solar filter can be tailor-made to your own specification. However the sheet material tends to degrade with use and after a while you have to replace it. Figure 8.2 shows a home made solar filter.

Figure 8.2. Home-made solar filter.

The Universe at a Touch of a Button

Figure 8.3. Glass solar filter for AR-6.

The color of the Sun through a Mylar filter normally exhibits a bluish white disc. Hence, the home made Mylar filter required a yellow plastic transparency to be added to the front of the solar film in order to colorize the Sun aesthetically yellow. The plastic sheet also acted to protect the solar film from accidental piercing. Glass filters however, normally provide a deep orange or pale yellow color.

Solar filters typically utilize the entire aperture of the telescope. However, for the some LXD telescopes models whose apertures are 8 inches in diameter or above, a reduced size filter is preferred. This is because the air currents inside the OTA are more turbulent during the daytime than at night and so a full aperture magnifies this turbulence. Therefore, by reducing the size of the aperture the focal length is increased thus reducing the effect of the turbulence. The AR-5 and AR-6 refractors are suitable for full aperture filters (Figure 8.3), whereas the other telescopes in the series such as the SC-8, SN-8 or SN-10, should have their apertures stopped down to around 5 or 6 inches, in order to get the best views of the Sun.

The filter normally fits snugly over the front of the telescope. However, as a precaution, it should be held firmly in place with either Velcro or insulating/duct tape. It is important to make sure that the filter does not fall off whilst the telescope is pointing towards the Sun! Even just a brief glimpse of the Sun through an eyepiece of an unfiltered telescope will cause irreparable damage to the eye. Figure 8.4 shows an image of the Sun through a home made solar filter.

Finding the Sun. Trying to get the Sun in the field of view of an eyepiece is actually not a trivial task. Obviously you cannot use the finderscope to visually sight the Sun, as it is just as dangerous as looking at it through the eyepiece of the main telescope. So what do you do?

For safety reasons the Autostar does not have the Sun in its object database, so you simply cannot Goto to it. You could try and work out what the Sun's current sidereal co-ordinates are, then set the telescope up so you can simply 'dial' its location, but that could take some time. The quickest and safest way, is to look at shadow of the telescope on the ground and note the shape of the OTA and connect accessories sticking out from the telescope such as the finderscope. Finding the Sun is as follows:

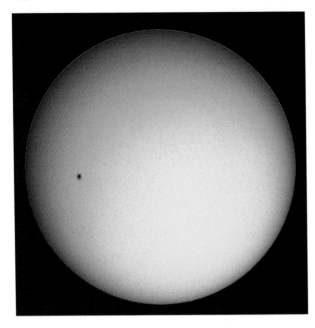

Figure 8.4. The Sun through a glass filter. Image courtesy of George Tarsoudis.

Before you start, make sure the solar filter is firmly fixed to the front of the OTA and that the front lens of the finderscope is covered. Orientate the OTA, such that the size of the shadow of the finderscope on the ground is at its smallest shape. This signifies that the finderscope is pointing directly at the Sun, and should be in the field of view of the eyepiece. There are now specially designed solar finders commercially available, which makes it much easier and safer to find the Sun with a telescope.

What Can You See on the Sun's Surface? So what can you see using white light solar filters? Well, quite a lot actually! The Sun is a dynamic body. It is constantly in a state of flux and you can see different features on a day by day, even hour by hour basis.

Through a white light solar filter the Sun's Photosphere is visible. It is a region some 6000 K degrees in temperature. Cooler regions, almost 2000 K less than the rest of the surface, known as *Sunspots*, are sporadically dotted around the equatorial region. Sunspot numbers vary over an eleven year cycle between peaks of solar activity. Even at solar minimum one or two Sunspots are still seen. At high magnifications, Sunspots reveal a structure; the *Umbra*, which is the central region and the surrounding *Penumbra*, which is lighter in color.

Lighter patches on the Sun's surface called *Faculae* can be seen. These are less obvious than Sunspots, and their visibility fluctuates according to the atmospheric seeing conditions. They often reveal themselves in moments of good seeing and being bright. They are often seen at the Sun's limb.

Solar filters allow us to see an effect known as *Limb Darkening*. The edge of the Sun appears to darken very slightly, compared to the central regions. This effect is due to the fact that light is being increasingly absorbed by material in the solar photosphere, as we look towards the edge of the Sun.

The Universe at a Touch of a Button

Solar Eclipses must be viewed using a solar filter during the partial stages. If you happen to be lucky enough to witness a total eclipse of the Sun, then you might be able to observe solar prominences on the Sun's limb with your telescope, during totality. You must take care though, as the event never lasts more than a few minutes, so you must immediately revert back to using a Solar filter once totality is over. You can use the Autostar to list forthcoming Solar eclipses (Total, Annular and Partial). The list is found under the 'Event' option.

`Select Item: Event → Solar Eclipses`

Autostar also provides Sunrise and Sunset information which depends upon your location and date settings. This is also found through the `Select Item: Event` menu option.

The planets Mercury and Venus transit the Sun's disc from time to time. These are seen as small circular discs against the background solar surface. I will talk more about transits later in this chapter.

The Moon – Lunar Observing

`Select Item: Object → Solar System → Moon`

Our nearest rocky neighbor, the Moon, is a fascinating place to explore with a telescope. Craters of all shapes and sizes are seen along with flat plains called *Maria*.

The Moon's phase changes over a period of around a month, and during that period different surface features are revealed. The linear area at which the separation of the light and dark areas of the surface is seen (known as the *terminator*), allows long shadows to be cast across crater floors from crater rims and peaks. This provides an almost three dimensional aspect to the lunar landscape. At full Moon however, most of the cratered topography appear flat. The Sun is shining perpendicular to the lunar surface so shadows are minimal. Bright rays and Maria are plainly seen during a Full Moon, and is still an interesting object to observe even during this period. At magnitude -12.5, the Full Moon is so bright that it will immediately destroy any dark adaptation you have as soon as you look at it through a telescope. Moon filters are available which help reduce the glare produced by the Moon.

The Moon's disc spans half a degree so by using a 26 mm eyepiece with any LXD telescope the whole of the disc is just able to fit in the field of view (Figure 8.5). Using a 9.7 mm eyepiece, you can carry out a closer inspection of lunar surface.

There are many lunar atlases available to help you navigate your way around the lunar surface, (See bibliography). Table 8.1 lists ten features that you might find of interest. Figure 8.5 shows where they are. Some are more challenging than others, and some are best seen when the Moon is at a particular phase of the lunar cycle. Autostar provides details of Moon Phases for any given date as well as when the Moon rises and sets. It also provides information about Lunar Eclipses (Total and Partial). This is done through the `Select Item: Event` option.

The impact basin Mare Orientale, the Eastern Sea, is a challenging feature to observe notably, because it is on the utmost Western edge of the Moon (yes! that's right Western not Eastern!). However, there are occasions when more of the impact basin can be observed due to *Libration* effects of the Moon. Libration is an irregularity in the Moon's

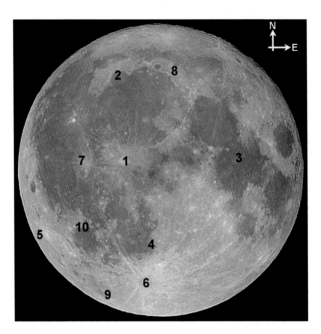

Figure 8.5. Ten interesting Moon features. Moon Image courtesy of George Tarsoudis.

Table 8.1. Lunar Features

Feature	Description
1	Copernicus (Crater)
2	Sinus Iridium (Bay)
3	Mare Tranquillitatis (Mare)
4	Rupus Rectus – Straight Wall (Ridge)
5	Mare Orientale (Impact Basin)
6	Tycho (Crater)
7	Kepler (Crater)
8	Vallis Alpes (Alpine Valley)
9	Schiller (Crater)
10	Gassendi (Crater)

rotation with respect to the Earth. This makes parts of the Moon's surface appear that would otherwise not be normally visible.

From time to time, Planets are occulted by the Moon. This can be a spectacular sight through the telescope, as you see the planet's disc pass slowly behind the lunar limb.

Mercury

Select Item: Object → Solar System → Mercury

The closest planet to the Sun is not an easy object to find as it is never far from the Sun in the sky. In fact not many people have seen it at all.

The Universe at a Touch of a Button

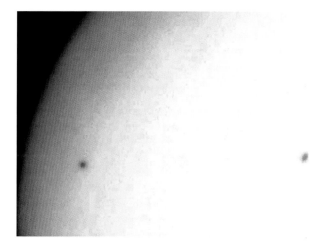

Figure 8.6. Mercury transit May 2003 (Mercury is on left, Sunspot shown on edge).

The best time to see Mercury is at greatest elongation, the point at which the planet is at its maximum angular distance from the Sun as viewed from the Earth. This is always at dusk or dawn.

You need to take precautions in finding Mercury with a telescope. You must wait until the Sun has completely dipped below the horizon before you can start looking for the Planet. This is so you don't accidentally glimpse the Sun through the eyepiece, when sweeping the sky for the Planet or, when the telescope is sweeping the sky whilst it is searching for the Planet.

Mercury undergoes phases just like the Moon. Through a 26 mm eyepiece Mercury appears as a small disc. Its phase depends upon its relative position with the Sun and the Earth. Details are not seen at higher powers, and the Planet tends to be difficult to keep in focus, due to its low proximity with the horizon.

From time to time, Mercury passes in front of the Sun as a transit. Through solar filters, the Planet can be seen as a tiny black disc against the bright background surface of the Sun. The last transit was in 2003 (Figure 8.6).

Venus

Select Item: Object → Solar System → Venus

The second closest planet to the Sun is an unmistakable object to see in the pre-dawn or evening twilight sky, shining at magnitude –4.3. The brightness is due to Venus's total cloud cover that reflects 72% of sunlight. Through an AR-5 or AR-6, Venus exhibits a false purple color around its limb. This is due to the achromatic optics of the telescope causing chromatic aberration effects. The effect is enhanced in particular when observing Venus, due to its brightness and small apparent size (approx. 1 minute of arc). Filters reduce the false color effect to an appreciative level. The SNT and SCT however, suffer no such color effect, due to being almost inherently achromatic.

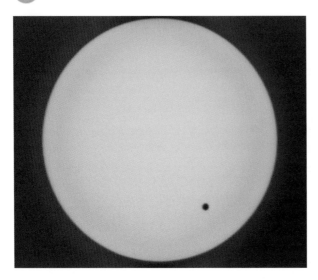

Figure 8.7. Venus transit image taken 8 June 2004.

Like Mercury, Venus shows phases and changes its apparent size during its orbit around the Sun. In visible light, the Planet is featureless and shows no obvious discernable markings.

Also like Mercury, Venus transits the Sun, although much less frequently (two events every 150 years or so). The last transit was in 2004 (Figure 8.7) and the next one will not be until 2012.

Earth

Of course there is no Earth entry in the Autostar database, however there is a Landmark facility which allows you to carry out terrestrial observing of the local scenery.

Terrestrial Landmarks.

Select Item: Object → Landmarks

Before you start observing terrestrial landmarks, you will need to disable the tracking feature of the mount, if it has not already been switched off. This will turn off the RA tracking motor, so the landmark can be viewed without the telescope slowly moving off its target. Scroll to the 'Targets' menu option and setting the target to 'Terrestrial'.

Select Item: Setup → Telescope → Targets → Terrestrial

By default Autostar does not have any Landmarks in its memory, so you will need to enter them into the handset. This is done by using the 'Landmark → Add' option. You will be prompted first to provide a name of the Landmark. Center the Landmark in the eyepiece using the Autostar arrow keys and press 'Enter'. Once the landmarks are stored in the Autostar database, you can call them up any time by scrolling down the user defined list. You can also perform a 'Landmark Survey', which is similar to the astronomical 'Guided Tour' but for terrestrial objects. You have to remember though, that to ensure the telescope is pointing to the correct landmark, it has to be

The Universe at a Touch of a Button

in the same position as it was, when the landmark was first entered into the database. Furthermore, when observing landmarks, there is a minimum usable distance before which a telescope can no longer physically focus onto an object. For objects closer than this limit you would have to extend the focusing tube in order to achieve a focus.

Observing Orbiting Earth Satellites.

Select Item: Object → Satellite

Sometimes when you are looking up at the night sky you may see a point-like object moving steadily across the sky. The object could be an aircraft at very high altitude, but more often than not, it is an Earth orbiting Satellite. You may also sometimes see these Satellites through an eyepiece as they move swiftly through the field of view. There are thousands of Satellites orbiting the Earth. Autostar contains by default, fifty Satellites including the Hubble Space Telescope and the International Space Station.

When a Satellite is selected from the Autostar database, (if it is forecasted to be observable) the AOS (Acquisition of Signal) and LOS (Loss of Signal) can be viewed, using the scroll keys. The difference between the AOS and the LOS is the length of time the Satellite will be visible above the horizon.

The Autostar will sound an alarm, which will denote that there is one minute left before the Satellite will become visible in the sky. You will need to navigate back to the Satellite list and select the same satellite again, before pressing the Goto button. The telescope will slew to where the satellite is predicted to be crossing the sky. A countdown timer will be displayed, and at about 10 seconds before the event takes place, you should prepare to look through the eyepiece. As soon as the Satellite is visible in the field of view, press 'Enter' and the telescope will attempt to track the satellite. You can center the Satellite in the field of view using the arrow keys without interfering with the tracking speed.

If a Satellite is not going to be visible at the time of the selection, the Autostar will display "*Calculating... No Passes Soon*", along with a beep sound. If the satellite information has not been updated recently (three to four weeks), then you may also get an "*Elements expired*" warning. To find out the status of the satellite information you should check the '*Epoch Year*' by selecting the 'Edit' option and then the appropriate satellite. You can also download new and updated satellite information into the handset, from Meade or other websites. (See Chapter 9).

Mars

Select Item: Object → Solar System → Mars

Through a telescope Mars appears as a reddish-orange disc. Its brightness (average magnitude -2.0) and apparent size varies distinctly over a period of a Martian year as the Planet journeys around the Sun. The maximum size also depends upon times of closest approach to the Earth, which occur every 2.5 years. The last major close approach was in August 2003. Many astronomers took advantage of the exceptional apparent size (25 arc sec) to observe the Planet in glorious detail. At other times Mars's apparent size during closest approach, varies between 15 and 20 arc sec.

At magnifications greater than $100\times$, features such as Polar ice caps and dark patches can be seen. The Polar ice caps shrink and grow over a Martian year and sometimes

Figure 8.8. The Planet Mars. Image courtesy of David Kolb.

the Southern ice cap disappears completely. Mars rotates just in over 24 hours. So, watching Mars at the same time over a period of several Earth days, you will be able to see different features rotating into view, traveling left to right across the Planet's disc (Figure 8.8). From time to time dusts storms take place obscuring the Planet's dark features, and sometimes envelope the entire planet.

Observing details on Mars very much depends upon local seeing conditions and the position of Mars in the sky. Observing the planet low down in the sky combined with poor seeing conditions, will wash out most of the features, leaving you with a relatively bland orange disc. The best way to observe Mars is to stare at it for a short period of time. During fleeting moments of atmospheric calm, the details will stand out. Red and Green filters sharpen the image and enhance the dark patches respectively. Blue and Violet filters improve the contrast between the polar ice caps and the rest of the features on the disc, and can even show clouds far above the Martian surface.

A really useful Mars observational tool can be found on the Sky and Telescope website called *Mars Profiler* (see Appendix D for website info). For any given date and time the profiler will display an annotated map of the Martian surface and provide observational characteristics such as visual magnitude and apparent size.

Jupiter

Select Item: Object → Solar System → Jupiter

Mighty planet Jupiter is easily recognized in the night sky shining at magnitude −2.7. Through a 26 mm eyepiece, the Gas Giant reveals a dynamic cloud band system with the most immediately obvious bands being either side of the equatorial belt (Figure 8.9). Jupiter spans almost 48 arc sec across at opposition, so a reasonable amount of detail can be seen on its disc, even at relatively moderate magnifications 100 to 200×.

At higher magnifications of 250× or more, the cloud bands show structure including white spots and filaments. One particularly large permanent feature in the Southern equatorial region, called the 'Great Red Spot', changes color over time. In recent years

The Universe at a Touch of a Button

Figure 8.9. The Planet Jupiter. Image courtesy of David Kolb.

it has not been as prominent, making it sometimes difficult to distinguish from the rest of the cloud belts.

Jupiter's fast rotation (under 10 hours) means that, in just a period of a few hours, you will be able to see significant movement of features across the disc, as other features rotate into view. You can also observe limb darkening similar to the Sun, but to a lesser degree.

The use of filters enhances details on the Jovian disc. A Blue filter is useful to enhance the Great Red Spot. Other colors such as Green and Orange filters enhance cloud belts and white spots on the disc.

Jupiter has a multitude of moons. Ganymede, Callisto, Europa and Io, are easily seen at low powers as bright star-like objects, never venturing far from the Jovian disc. Their orbital motions are fascinating to watch as they transit, eclipse and occult their parent planet and each other. These events can be predicted well in advance. Details can be found in most astronomy magazines and the Internet.

Saturn

Select Item: Object → Solar System → Saturn

Saturn is an unmistakable object in a telescope shining at magnitude +0.7. Through a 26 mm eyepiece Saturn's disc is much smaller than Jupiter's (19 arc-sec at opposition), but the planet has a significant ring system. During times when they are fully open, the Planet's overall image size is comparable to that of Jupiter's (45 arc sec).

The ring system is split up into a number of designations from the outer- most prominent ring A, to G much further in. There are also gaps between the rings. The two most prominent gaps are the Cassini and Encke divisions. The Cassini division lies between rings A and B, and is likely to be the most immediate feature you can see when you first observe the rings. The Encke division lies just outside the Cassini division and requires exceptional seeing conditions and high magnifications to view it. Further in towards Saturn, lies the Crepe or C ring, which appears as a dark almost transparent-like ring on the inner edge of the B ring.

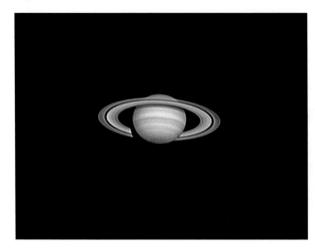

Figure 8.10. The Planet Saturn. Image courtesy of David Kolb.

Like Jupiter, Saturn also has cloud band systems and occasional storms seen as white spots, although they are apparent to a lesser degree. Filters used such as Yellow and Red filters enhance the cloud belts on the planetary disc whilst a Green filter highlights the rings. In any case Saturn is a beautiful image to observe (Figure 8.10) and always prompts the 'Wow' factor whenever it is shown to people who see it for the very first time.

Saturn has a large number of orbiting Moons. Once such moon, Titan, has a significant atmosphere and is easily seen in telescopes at low powers, around magnitude +8.

Uranus, Neptune and Pluto

Select Item: Object → Solar System → Uranus

Select Item: Object → Solar System → Neptune

Select Item: Object → Solar System → Pluto

Far beyond Saturn lie the two gas giants Uranus and Neptune and the remote icy body Pluto. *Uranus* shines at magnitude 5.5, which means it is just visible to the naked eye. Through a telescope with moderate magnifications (100× or more), Uranus appears as a small bluish-green disc only 3.9 arc-sec across at times of opposition. Even at high magnifications (greater than 200×), the disc is bland and featureless with no discernable belts or structure. (Figure 8.11). Uranus has moons of its own however, you will need to use the largest of the apertures, such as the SC-8, SN-8 or SN-10, in order to reach the faint magnitudes these satellites are visibly seen (+13.5 to +14.5).

Neptune is even more distant shining at magnitude +7.8. In telescopes Neptune appears as a very small dark-bluish disc just 2.4 arc sec across, this is almost 20 times smaller than Jupiter's disc. Even at very high magnifications 300× or more you will struggle to notice any details on the planet's disc.

The Universe at a Touch of a Button

Figure 8.11. The Planet Uranus. Image courtesy of David Kolb.

Pluto is a challenge to find. The planet's remoteness from the Sun and low brightness (+13.9), means that it is only seen as a point-like object in even the largest of the LXD telescopes SC-8, SN-8 or SN-10. The only true way to find Pluto is with detailed finder-charts, and by looking in the area where the planet is located over a period of a few nights. You will see a star-like object appear to move against the background of stars. Of course, since the IAU has demoted this little body to a 'dwarf planet', then you may see it eventually demoted to a different object list!.

Asteroids

Select Item: Object → Solar System → Asteroids

Lying primarily between Mars and Jupiter, the Asteroids, also known as Minor Planets, constitute a group of small bodies made up of rock, iron and other compounds.

The brightest asteroids can regularly be seen wandering amongst the stars. In telescopes, asteroids are seen as point like objects and are best observed over consecutive nights in order to determine their path of motion. Table 8.2 shows a list of ten asteroids

Table 8.2. Ten Asteroids to Find

Asteroid Name	Visual Magnitude
1 Ceres	+7.8
2 Pallas	+7.6
4 Vesta	+7.3
6 Hebe	+8.8
7 Iris	+7.0
11 Parthenope	+11.2
14 Irene	+10.9
18 Melpomene	+9.9
29 Amphitrite	+9.5
349 Dembowska	+10.2

that you can find using your telescope. The numbers next to their names are the order of discovery. Some may be more challenging to find than others.

You can manually add, modify or delete asteroid information in the database. To add a new asteroid to the database, you will need to know its orbital elements, which you should be able to find from various astronomical data references. You can also download new asteroid information electronically via the Meade website.

Comets

Select Item: Object → Solar System → Comets

Comets are the fleeting travelers of the Solar System. They are made up of ice, rock and dust. Long-term periodic Comets orbit the Sun out to Neptune and beyond, such as Halley's Comet which comes round once every 76 years. Short-term periodic Comets such as Encke or Wild reside within the confines of the inner solar system. They are frequently observed but most are not very spectacular to look at, appearing as fuzzy faint blobs in the eyepiece. Sometimes a Comet comes from the remote outskirts of the Solar System and makes a spectacular appearance, a once in a lifetime event not to be missed.

Comets that exhibit distinct Ion and Plasma tails are best seen with wide-field, low magnification eyepieces. Sometimes the tails are so long that a pair of binoculars is the preferred instrument, to view the Comet in all its glory. For bright Comets and those that pass close to the Earth, higher magnifications may reveal structure in the central nucleus.

By default, Autostar has a list of only fifteen Comets. You can add new ones by manually entering in the comet's orbital data or electronically download from the Meade website.

Events – Meteor Showers

Select Item: Events → Meteor Showers

When you are out observing, you will sometimes see a shooting star or Meteor. The Meteor may be sporadic or it may form as part of a periodic meteor shower. Meteor showers are formed when the Earth passes through a dust stream expelled by a comet on its journey around the Sun.

Meteor showers appear at regular intervals throughout the year, some are more exciting than others. Appendix B provides a list of annual meteor showers, as stored in the Autostar database. The name of the shower signifies a point in the constellation for which the shower originated from (the radiant). It is best to observe Meteors about 45° either side of the radiant, in order to get the maximum effect.

The position of the Moon plays an important part as to whether a meteor shower is seen or not. A Full Moon tends to make observing meteors showers difficult. The sky is brightened to such a degree, that it drowns out all but the very brightest of Meteors.

If so desired you can direct the telescope toward the radiant by pressing 'Goto' on the particular Meteor shower selection. However, the telescope's field of view even at its maximum will not be suitable to see an entire display, so it is best to witness the event with no instruments at all, using just the naked eye.

Constellations

> Select Item: Object → Constellations

There are 88 constellations in the sky. Selecting a constellation with the Autostar, then pressing the 'Goto' key, displays a list of brightest stars in that constellation. Scrolling down the list of stars pressing 'Goto' again, and the telescope will slew to that star. This useful feature means that you can tour the stars of any constellation in the sky. A list of all constellations is in Appendix B.

Stars

> Select Item: Object → Star

Through a telescope, stars are seen as point-like objects. Many astronomers use them merely as pointers to search for more interesting objects such as Galaxies and Nebulae. However, stars have interesting characteristics of their own and can be just as visually rewarding as with other objects.

Stars are of all different types, some vary in brightness in a predictable way or have irregular variability. Some stars are in double or multiple systems with two or more orbiting one another. They vary in size, with color directly related to their temperature. I talked briefly about stellar classification in Chapter 2.

Autostar has literally thousands of stars in its database taken from the Harvard Bright Star Catalog and the Smithsonian Astrophysical Observatory (SAO). The 'Star' menu is split into several options (see following sections).

Named Stars

> Select Item: Object → Star → Named

Many stars have proper names rather than Greek/Bayer letter designations. The names usually have some significant Latin or Arabic meaning based on legend or myth. Some stars are even named after their unusual visual properties such as the Demon Star. The Autostar has 79 named stars in its database. They are used in preference to their Greek designations as specified in other star lists elsewhere in the database.

The named star database is also used for Goto setup star alignment and a list along with finder charts are in Appendix A.

Double Stars

> Select Item: Object → Star → Double

When you look at a star through an eyepiece, you may be surprised to find that it is not a single entity at all, but is split up into two or more separate entities. There are two types of these stars; those that are gravitationally bound in the same star system known as Binary Stars, and those that are in close vicinity to each other in the sky through line of sight only. These bear no physical influence upon each other and are known as Optical Doubles or Optical Binaries.

Some binary star systems are such that one star eclipses the other one, therefore affects the light output from the entire system. These are Eclipsing Binaries. A good example of an eclipsing binary is 'Algol' (Beta Persei). Over a period of almost three days, Algol A, the primary star in the system, dims for a few hours as its companion star Algol B passes in front of it (there is actually also Algol C). Algol A varies between magnitude 3.5 and 2.3 at times of minimum. You can find out when the next minimum of Algol occurs using the Autostar, but it depends upon the date you entered when the Autostar was first powered on. To predict subsequent minimums of Algol you will need to change the date of the handset.

> Select Item: Event → Min. Of Algol

In Chapter 7, I suggested that a telescope's ability to resolve a double or binary star depends primarily upon the size of its aperture. Observing double stars through the telescope is a good test of its optics and the local seeing conditions, especially if the stars are almost at the resolution limit of the telescope.

Variable Stars

> Select Item: Object → Star → Variable

Most stars live on the Main sequence, a stage in their lives when their energy output does not fluctuate over millions of years. Other stars vary their brightness over much shorter periods, years and even months.

There are various reasons why a star undergoes variability. It could be that it has not long formed or it is coming to the end of its life, i.e. it is unstable. Other stars undergo more powerful changes. These flare up thousands of times their normal intensity for just a few weeks, before they revert back to their normal selves again. They are termed 'Novae' and there are known mechanisms that cause the regular flare-ups.

Some amateur astronomers dedicate themselves to the study of variable stars in great detail, in order to determine the physical mechanism behind their variability. In fact, a particular group of periodic variable stars, called Cepheids are used to determine the distance to distant Galaxies.

You too can carry out variable star observations, although it does takes some practice to estimate a star's brightness to an accuracy of a tenth of a magnitude, or better. The variable star list in the Autostar is extensive and quite long, containing around 180 or so entries. If you know which variable you need, then unfortunately you may have to scroll past tens of entries until you find it. This could take some time, as well as suffering from an achy thumb from pressing the scroll keys many times! If you know the RA and DEC of the variable star, then you can shortcut this list and create a new user object (see later in the chapter), or add it to a customized guided tour.

Other Star Catalogs

SAO Star Catalog.

> Select Item: Object → Star → SAO Catalog

The Autostar SAO catalog list is a cut down version of the real SAO catalog, where the original catalog contains 258,996 stars down to magnitude +11.0.

The SAO star catalog boasts around half the number of objects in the Autostar database, 18,737 in total. So, unless you know the SAO number for a particular star, you may spend many hours (or waste many hours depending upon your point of view), entering random five or six digit numbers into the handset and see if the star exists or not... then again you might want to try out this method one night and see where it takes you in the sky.

Nearby Stars.

`Select Item: Object → Star → Nearby`

There are 23 objects of nearby stars in the Autostar list. The list includes Rigel Kentaurus whose name is more commonly known as Alpha Centauri, the closest star to our Solar System.

With Planets.

`Select Item: Object → Star → With Planets`

We live in an age when extra-solar planets have been discovered around other stars. Although, astronomers are yet to determine the actual properties of these planets, more are being discovered every year. The Autostar lists a small selection of relatively bright stars that were recently discovered to have planets orbiting them. The list tends to provide the SAO designation of the star rather than the name for which it is commonly known. Three stars in the list known to have planets around them are 51 Pegasi (SAO 90896), 47 Ursae Majoris (SAO 43557) and 70 Virginis (SAO100582).

Deep Sky

`Select Item: Object → Deep Sky`

The Deep Sky object menu is extensive and there are over 13,000 objects to choose from. It is divided into various categories, described in the following sections. Some of the objects in the lists, such as the Andromeda Galaxy or Orion Nebula are bright enough to be easily visible with the naked eye. Others, such as the Black Hole Cygnus X-1 at magnitude 20.0, are too faint to be seen visually even through the largest LXD telescope model, the SN-10.

Object Catalogs

The Autostar has several catalogs in its database. These catalogs are well known and are the basis upon which all amateur astronomers identify objects. Some objects are recognized via a common name e.g. Wild Duck Cluster, Whirlpool Galaxy, Swan Nebula and so forth. Many deep sky objects are referenced in more than one object catalog and therefore have multiple designations. For instance, the Andromeda galaxy is listed in the Named Objects list and Galaxies list, Messier list as M31 and the NGC list as NGC 224.

The Messier Catalog.

Select Item: Object → Deep Sky → Messier Objects

Named after Charles Messier who first compiled the catalog in the late 1781 to determine which objects in the sky were not Comets. The Messier catalog primarily consists of objects located in the Northern hemisphere. All objects in the Messier catalog are identified with a prefix letter 'M'. A complete list of Messier objects is in Appendix B.

NGC Objects.

Select Item: Object → Deep Sky → NGC Objects

The New General Catalog (NGC) is the most popular list of deep sky objects used by amateur astronomers. The catalog contains different types of deep sky objects such as Galaxies, Nebulae and Star Clusters, and incorporates both the Messier and Caldwell catalogs (see next section). A list of NGC objects is not provided in this book as there are 7840 in total.

Caldwell Objects.

Select Item: Object → Deep Sky → Caldwell Objects

The Caldwell Catalog is a list of 109 objects conceived by the astronomer Sir Patrick Moore. The catalog is intended to supplement the Messier catalog, hence there is no equivalent Messier designation. All objects in the Caldwell Catalog are identified with a prefix letter C. A complete list of Caldwell objects is in Appendix B.

IC Objects.

Select Item: Object → Deep Sky → IC Objects

Where the Caldwell catalog serves as a supplement to the Messier Catalog, the Index Catalog (IC) serves as a supplement to the NGC objects list. The IC list is too large to list in this book (5386 objects).

Named Objects

Select Item: Object → Deep Sky → Named Objects

The Named Objects deep sky list is a mixed collection of Galaxies, Nebulae, star clusters including exotic objects such as Quasars and Black holes.

Most objects in the list have common names astronomers recognize, such as the Pleiades, Dumbbell Nebula (Figure 8.12) or the Whirlpool Galaxy. Other objects are simply designated by long sequences of numbers and letters e.g. PKS0405-12. It is mystifying to suggest why these objects are in the named objects list, since they are not really classed as 'named' objects.

The Universe at a Touch of a Button

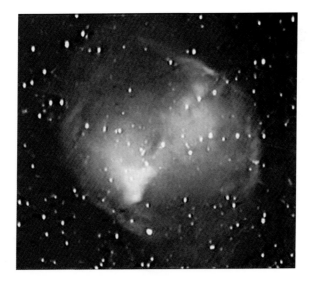

Figure 8.12. M27 Dumbbell Nebula in Vulpecula. Image courtesy of Dieter Wolf.

Galaxies

Select Item: Object → Deep Sky → Galaxies

The Universe is full of Galaxies millions if not billions of light years away from our own Galaxy the Milky Way. The light that left these remote objects has taken millions of years to reach us, so we are essentially looking back in time at them to what they looked like millions of years ago.

The most suitable time of year to observe them is during the spring where the constellations Coma Berenices, Leo and Virgo are visible in the sky. Galaxies congregate in clusters and superclusters. In the vicinity of the Virgo region a large supercluster exists which contains thousands of faint Galaxies. In fact our Milky Way is a member of this particular supercluster. Sweep the Virgo area with your telescope in dark skies, you might pick up one or two in your eyepiece.

The SN-8, SN-10 and SC-8 have the light grasp to detect faint Galaxies down to 13th magnitude or so. However, some Galaxies may be listed as brighter than 13th magnitude but this is the integrated magnitude and not its true brightness. Hence, you may find that a galaxy's apparent visual brightness is a few magnitudes fainter than specified in the Autostar database. Viewed through these telescopes, Galaxies are seen as fuzzy misty patches of varying shapes. In fact, some are so faint that they almost merge into the background sky, so the best way to observe them is to use averted vision (see Chapter 7).

If you wish to use a higher power to observe a Galaxy such as a 9 mm or 12 mm eyepiece, you will find that it will be larger in the field of view, but will appear fainter, so Galaxies are best viewed in moderate to low power eyepieces Figure 8.13 and Figure 8.14 show the Whirlpool Galaxy and the large Galaxy M82 respectively.

Figure 8.13. M51 Whirlpool Galaxy in Canes Venatici. Image courtesy of Dieter Wolf.

Figure 8.14. M82 Galaxy in Ursa Major. Image courtesy of Dieter Wolf.

The Universe at a Touch of a Button

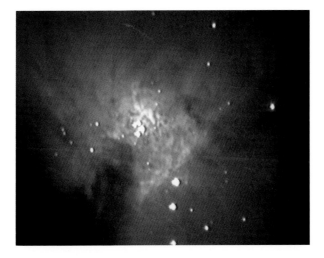

Figure 8.15. Central region of M42 with trapezium. Image courtesy of Dieter Wolf.

Nebulas

Select Item: Object → Deep Sky → Nebulas

Through a telescope Nebulae appear as misty patches. With the exception of a few bright ones, most visually lack any distinct color unless they are imaged. Many are named after their distinct shape, e.g. North American Nebula, Eagle Nebula, Owl Nebula etc.

There are various different types of Nebulae; Dark, Diffuse and Planetary. Some are the remnants of exploded stars. Diffuse Nebulae are of two types; *Emission Nebulae*, where they emit light from within the Nebula through internal processes e.g. the Orion Nebula (M42) (Figure 8.15), and *Reflection Nebulae*, where light is reflected from nearby stars e.g. NGC 1435 in the Pleiades. Some Nebulae are a combination of both types like the Trifid Nebula (M20) in Sagittarius.

Select Item: Object → Deep Sky → Planetary Neb

Planetary Nebulae, as their name suggests are disc-like in shape. Some closely mimic a planet's shape like the Saturn Nebula (NGC 7009) in Aquarius. Probably the two most often observed Planetary Nebulae are M57 the Ring Nebula in Lyra and M27 the Dumbbell Nebula in Vulpecula. Both are very different looking objects; through a 26 mm eyepiece M57 looks like a small smoke ring (Figure 8.16), whilst M27 has a 'dumbbell' or 'apple core' shape. The Autostar has several different types of Nebulae including Planetary Nebulae in its database.

Dark Nebulae are not internally illuminated nor do they reflect light from other sources. They are seen as dark patches in front of stars or diffuse Nebulae e.g. Horsehead and Coalsack Nebulae.

Filters such as Lumicon's Ultra High Contrast are used to bring out details in most Nebulae and Planetary Nebulae.

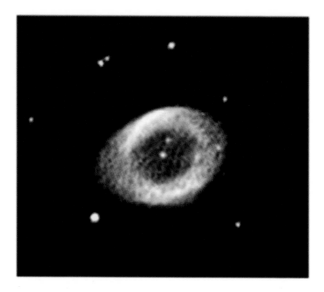

Figure 8.16. M57 Ring Nebula in Lyra. Image courtesy of Dieter Wolf.

Star Clusters

Select Item: Object → Deep Sky → Star Clusters

Many stars in the sky are seen in groups. There are different types of star clusters, ranging from loose open clusters of perhaps a few tens of stars, to tightly packed globular clusters containing thousands or even millions of stars.

It is best to view star clusters in low power wide field eyepieces, in order to appreciate their splendor. However, in cases such as the open clusters, they may be too wide to fit within the field of view, and only sections of the cluster are seen at any one time.

The Star Clusters Autostar list is a mixed list of Messier, Caldwell and NGC objects. Named objects include the Pleiades (M45) and Hyades in Taurus, Praesepe – the 'Beehive' (M44) in Cancer, the double cluster in Perseus (C14), M3 (NGC5272) in Canes Venatici (Figure 8.17) and Omega Centauri in Centaurus (NGC 5139).

Exotic Objects

Select Item: Object → Deep Sky → Quasars

Select Item: Object → Deep Sky → Black Holes

Quasars and Black Holes are enigmatic objects. Most of the objects in the Autostar lists specify magnitudes that are too faint to be seen with any LXD telescope. The exception is the Quasar 3C273 in Virgo which shines at magnitude +12.8 and is the brightest Quasar in the sky. Through an eyepiece it is seen as a star like object. 3C273 is in fact over two billion light years away, so the next time someone asks you how far you can see with your telescope, mention this interesting little fact!

The Universe at a Touch of a Button

Figure 8.17. M3 (NGC 5272) in Canes Venatici. Image courtesy of Dieter Wolf.

Using Autostar to Identify Objects

The Identify function is a useful tool to help you find you way around, if you are not sure the object you are currently observing, or even where you are pointing the telescope in the sky. The function works best when the telescope is accurately aligned with the Celestial Pole, and only if you are confident the Autostar will correctly identify the object. The steps to identify an object are as follows:

1. Point the telescope to an object or area of the sky you want to identify.
2. Go to the option Select Item: Object → Identify
3. The Autostar will display '*Searching...*'. After a few moments the handset will beep and the object description will be displayed.

If you are not pointing to any object in particular, then the nearest object is identified by the Autostar instead. You can then Goto the object should you so wish.

Browsing the Autostar Object Database

Select Item: Object → Browse

The Browse feature aids you in finding objects in the database without manually searching through thousands of objects. The browse utility however, is limited to deep sky objects such as Galaxies, Star Clusters and Nebulae.

Before you start a search of the database, you need to set up the search parameters. This is done by editing the parameters.

Select Item: Object → Browse → Edit Parameters

When the option is selected the search parameters will be listed:

−Largest (mins)
−Smallest (mins)
−Brightest (mag)
−Faintest (mag)
−Min El. (deg)
−Object Types

The 'Largest' and 'Smallest' parameters denote the object's apparent size in arcminutes. The 'Brightest' and 'Faintest' parameters denote the magnitude. 'Min Elevation' denotes the minimum height the object should be above the horizon. Object types are split into twelve subdivisions:

−Black Hole, Diffuse Nebula, Dark Nebula
−Asterism, Elliptical Galaxy, Globular Cluster
−Irregular Galaxy, Open Cluster, Planetary Nebula
−Quasar, Spiral Galaxy, Unnamed

Say for instance, you want to find all Galaxies in the sky at the current time, with a magnitude range of between +9.0 and +14.0:

1. Set the Largest value to default value 250.
2. Set the Smallest value to 0.
3. Set the Brightest value to magnitude +9.0 (use the left arrow key to change the plus/minus magnitude sign).
4. Set the Faintest value to magnitude +14.0.
5. Set the Elevation angle to 25°.
6. Scroll through the Objects types and use the 'Enter' key to select the type of object to search for, in this case set the 'Spiral Galaxy' and 'Galaxy' options. The plus and minus signs denote whether the object should be included or not. You can select more than one object type to search for.
7. Use the 'Mode' key to exit from the edit parameters option and go to the Start Search option.

Select Item: Object → Browse → Start Search

8. Select the *Next* option to start the search.
9. After a few moments the first object with the search parameters is displayed. You can scroll through the list by selecting either the *Next* or *Previous* options.

The Universe at a Touch of a Button

User Object

Select Item: Object → User Object

If an object is not in the Autostar database then you can add it to the custom list. Objects can also be added through the Autostar suite (see Chapter 9).

Sometimes you don't want to waste valuable observing time, searching through long lists of objects just to find the one you want e.g. a variable star. The User Objects function is very handy in this case, by allowing you to create a 'favorites' list, so you can quickly find the relevant object.

There are four options to manage user objects:

–Select
–Add
–Delete
–Edit

Add a user object: Select the 'Add' option, enter the Name, RA, Dec, Size, Magnitude of the object.

Select a user object: Choose the 'Select' option, use the up/down scroll keys to scroll through the list of objects.

Delete a user object: Select the 'Delete' option, use the up/down scroll keys to scroll through the list of objects. Pressing 'Enter' will delete the object listed.

Edit a user object: Select the 'Edit' option, use the up/down scroll keys to scroll through the list of objects. Pressing 'Enter', will list the properties of the object for editing Name, RA, Dec etc… in sequential order.

Suggest

Select Item: Utilities → Eyepiece Calc. → Suggest

This useful little utility suggests which eyepiece you should use, when you are observing an object selected from the Autostar database. The suggestion is based upon the telescope model and focal length data that was entered into the Autostar database. Of course, you may not own the eyepiece that the Autostar has suggested, but it gives you an idea of what should be the best one to use.

Summary

I hope that this chapter has given you just a taste of what is out there in our Universe. There are so many objects in the sky that you will never see all of them in a single lifetime, but by using a Goto telescope, you will be able to observe a good selection of wonders that the sky has to offer.

CHAPTER NINE

Connecting to a Personal Computer

Introduction

Nowadays most people own a personal computer. Telescope manufacturers have acknowledged that fact by building telescopes that have the ability to interface with a PC. Computerizing your LXD telescope means you can access and download objects to the Autostar, as well as control your telescope from a remote location.

There are many Astronomical Planetarium software packages available, which allow you to control your LXD telescope directly through the PC. A list of packages is in appendix D. One such package is the Meade Autostar Suite that comes with the LPI and DSI digital imaging cameras. The software package allows you to have complete control, as if you were using the Autostar handset.

Furthermore, the Autostar's internal software (firmware) can be updated through a PC connection using the Autostar Software Update (ASU) utility.

Connecting the Autostar to a PC

The Autostar connects to a PC via an RS-232 connection. RS-232 is the IT industry standard for serial data transfer between electronic devices. It has been superseded recently by faster and power efficient connection methods e.g. USB and wireless Bluetooth. However the current Autostar handsets in use have yet to catch up with the technology.

Figure 9.1. Connector ports on the Autostar.

The Autostar Serial Port

The serial port located at the base of the handset is a standard RJ-11 socket. This is smaller than the other ports, as it has only four pins compared to the other sockets' eight pins (Figure 9.1).

A PC on the other hand has a completely different type of serial port connection (9-pin D sub type male), so you will need to have a dedicated cable (RJ-11 to 9-Pin D sub-type female) in order to connect the two devices together.

Commercially manufactured cables are available through Meade, notably the #505 connector cable set (Figure 9.2). The cable sets are also included with the Lunar Planetary and Deep Space Imagers.

The standard commercial cable set comprises of two separate elements, which you connect together to create the complete cable. Cables exist, which are single element in design. However, these are normally of home-made construction, although in most cases they are made to a high quality standard. If you have some soldering skills, you can construct a cable yourself out of the necessary components available from any electronic component manufacturer (Radio Shack etc...). Wiring instructions for creating a serial cable can be found on Mike Weasner's Mighty ETX website (see Appendix D).

Figure 9.2. Meade #505 cable set.

Connecting to a Personal Computer

Figure 9.3. USB to serial cable.

Unless the telescope is permanently housed in an observatory, most astronomers tend to use a portable laptop for computer control capability out in the field. This is instead of hauling a bulky PC around the place. Most laptops nowadays do not have a serial port as part of their specifications, instead they have USB connections. There are commercially available USB-to-Serial adaptors, which essentially acts as a serial COM port running from a single USB port. This is the only method for connecting an Autostar to a PC or a Laptop that does not have a built-in serial port (Figure 9.3).

Computer Setup

The computer's serial port has to be configured to the correct speed for data transfer to and from the Autostar. For most Windows^{TM} based operating systems the settings for the serial port are found in the *Device Manager* drop down list from the main *Control Panel* option; usually listed as '*Ports (COM & LPT)*'.

By default the COM channel number for the serial port is COM1, however if you are using a Serial to USB adaptor, you may have configured the COM channel to a different channel Id, i.e. COM2. This should be listed along with other configured ports such as a printer port. The computer's serial port settings should be configured as shown in Table 9.1.

Table 9.1. Computer Serial Port Settings

Port Setting	Value
Baud Rate	9,600
Data Bits	8
Parity	None
Stop Bit	1
Flow Control	None

These settings are required otherwise problems might be encountered relating with the communication and data transfer between Autostar and the computer.

Upgrading the Firmware

The software on the Autostar can be upgraded. This means that new features and bug fixes can be uploaded to the handset.

To update the Autostar follow the steps below:

1. Ensure that the serial lead is connected to both the Autostar and the computer. The Autostar must be powered off first before you connect the serial cable to the serial port.
2. Power on Autostar. Enter the date and time.
3. Start up the ASU program on the computer. You will see a screen like the one shown in Figure 9.4.
4. Click on the "Upgrade Autostar Software Now" button.
5. After a few moments you will get a warning window on the computer screen with the following message "*Is Autostar in Safe Mode? (Does the display read "Flash Load Ready?)*". The Autostar #497 model screen actually displays "Downloading. Do not Turn Off", so click on the 'No' button. Please note this message is not displayed every time.

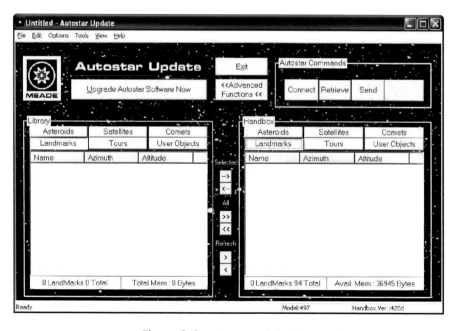

Figure 9.4. Autostar update utility.

Connecting to a Personal Computer

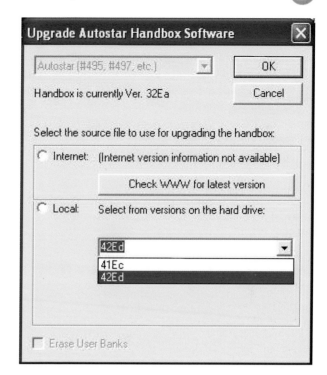

Figure 9.5. Upgrade Autostar handbox software window.

6. The "Upgrade Autostar Handbox Software" window is then displayed on the computer screen (Figure 9.5). The current version of the Autostar firmware is listed in the window.
7. You can choose two methods for upgrading the firmware; the internet and from the local hard drive.
   a. If you choose the internet, click on the "Check WWW for Latest Version" button. As long as you have a connection with the internet the ASU will automatically download the latest file.
   b. If you choose local hard drive then the ASU will search the 'Ephemerides' folder for suitable files and list them in the dropdown box in the window (Figure 9.5). See later for how to download from the internet without using the ASU.
8. When either choice from step 7 is selected, a window "Are you sure you want to upgrade now?" will be displayed. This is your last chance before the handset firmware is upgraded, and there is no turning back until the upload is complete, so think carefully before you click on the 'Yes' button.
9. A window '*Clear User Data. User Objects, Backlash Training, User Sites and Motor calibration will be retained. Use the <SETUP> - <RESET> from the handbox to reset.*" Will be displayed. Click 'OK' to continue.
10. Uploading to the Autostar commences. You can check the percentage status of the upload anytime by looking in the bottom left hand corner of the ASU window. Uploading the firmware file can take between 15 and 25 minutes. In the meantime,

during the upload process DO NOT power of the handset as you will corrupt the firmware being uploaded to the handset.
11. When upload has completed the Autostar will reboot and the computer will display the window "Data successfully sent to Autostar".
12. You can check the version of the new firmware by noting the version number when the Autostar firsts initializes or alternatively you can check the version from the Statistics option

> Select Item: Setup → Statistics

The ASU has a facility that allows you to backup data from the Autostar handset and store the information in a file on the computer. If there is a problem with the firmware you can easily revert back to the original version by simply reloading the previous version or restoring data.

Downloading Firmware Files from the Internet

Go to the Meade website (www.meade.com) and download the latest ASU and firmware file. The download file is in the form of a compressed Zip file, which contains the binary firmware file and a *readme* text file explaining the latest features and bug fixes included in the new firmware file.

Download the zip file into the *Ephemerides* folder, located in the ASU program area on the computer hard disk. The location should be something like *C:\Program Files\Meade\AutostarSuite\Updater\Ephemerides*. Unzip the contents of the zip file into the folder.

The firmware file is of the form Build*nnLL*.rom. Where *nn* represents the build number and *LL* denotes the minor update version of the build, e.g. Build42Ed.rom. Check the website regularly for updates and new Autostar information.

Other Features of the ASU

The ASU also allows you to edit, delete and create object data on the Autostar database, namely Asteroids, Comets, Landmarks, User Objects, Guided Tours and Satellites. Much of the information can be downloaded from the Meade website.

Autostar Cloning

Cloning is a useful method for transferring object catalog data between two Autostars without the need for a computer. You can even update the firmware from one Autostar to another.

Two Autostars are connected to each other through their RS-232 ports. One Autostar becomes the sender whilst the other one is the receiver. Each Autostar requires their own power therefore you will need two telescopes to carry out the cloning.

Connecting to a Personal Computer

Perform the following steps to clone an Autostar.

1. Connect the RS-232 lead to each serial port on both Autostar handsets. For simplicity lets call the sending Autostar '*As1*' and the receiving Autostar '*Ar2*'
2. Connect both Autostar handsets to their associated telescopes and power them on. Note: There is no preference to which handset you power on first.
3. On Autostar *Ar2* receiving the data go to the Download option:

   Select Item: Setup → Download

4. When 'Download' is selected the Autostar displays the message "*To Load Software. Press Enter*".
5. Pressing Enter will display the message "*Downloading. Do Not Turn Off*". The Autostar is now ready to receive data.
6. On the Autostar *As1* sending the data go to the Clone option:

   Select Item: Setup → Clone

7. Select one of three options to send to Autostar *Ar2*:
   Catalogs – sends user defined objects such as new comet data.
   Software – upgrades the receiving Autostar with the same firmware version as its own. This takes about as long as uploading the firmware through a computer.
   All – sends all information (catalogs and software).
8. When data has been transferred Autostar *Ar2* will sound a beep and reboot whereas the Autostar *As1* will revert to the previous clone menu.

Connecting Autostar to Other Devices

The serial interface on the Autostar does not totally confine it just to your PC or laptop. Any portable equipment with a compatible serial interface has the potential to connect to it.

Personal Data Assistants and organizers now have the processing capacity to control a telescope through the serial interface. Cables are similar to the #505 serial lead and are available from most telescope stockists.

CHAPTER TEN

Taking Images

Sometimes you will want to record the image of an object you observe through the telescope. Simply placing your camera up to the lens of the eyepiece and taking a few snaps, may work for some of the brightest of celestial objects like the Moon or the Planets. But to capture faint Galaxies or Nebulae requires a more dedicated imaging setup. Taking breathtaking shots of deep sky objects requires a lot of hard work and dedication. In the end you will have something that you can be proud to show to your friends and colleagues.

There are several methods for taking images of the night sky. One is the traditional tried and tested method of using a 35 mm camera and the other is through the use of digital cameras. Both are sound methods for capturing high quality images and pictures. With the advent of the digital age, the traditional 35 mm approach is rapidly giving way to digital imaging.

The aim of this chapter is to provide you with a brief insight into what you can achieve with your LXD telescope and appropriate imaging equipment.

Traditional Astrophotography

The following sections describing traditional equipment used to take astronomical photographs, notably mechanical cameras.

Choosing the Right Equipment

The best equipment suitable for taking photographs of objects in the night sky is a 35 mm SLR (Single Lens Reflex) camera with a 'B' (Bulb) time exposure setting. The

SLR feature allows you to view objects in the viewfinder as they are seen through the camera lens. The 'B' setting customs the time exposure such that the shutter will stay open for as long as the shutter release button is pressed down. It is best to use a camera with a 'B' setting that is mechanical and does not rely on battery power, otherwise you may find that whilst attempting a long exposure, especially in the cold, the battery will drain rapidly and the shutter will inadvertently release, cutting short the exposure time.

Some cameras also have the capability to replace the internal focusing screen with special screens in order to see objects much more clearly in the viewfinder. These screens are also designed to allow you to focus the object in the camera's viewfinder, much more easily than if a standard focusing screen was used. I own an Olympus OM-1 camera, which has a #1-8 Astro focusing screen and is ideally suited for astrophotography.

Reducing Camera Shake

Using your finger to hold the shutter button down is not very practical, especially if you are carrying out long guided exposures. A cable release is preferable in order to minimize camera shake. The cable release usually has a locking feature which locks the shutter release button in place, so you don't have to hold the cable during the full length of the exposure. It is best not to use a compressed air release cable for long exposure astrophotography, as the air tends to slowly leak away, accidentally releasing the shutter before the exposure is completed.

For SLRs when the shutter is opened, the mirror inside the camera flips up causing the camera to shake, in some cases enough to ruin the image. Most of the well-known camera brands, such as Olympus, Nikon, Canon and Pentax have the facility to lock the mirror to eliminate the shake before the shutter is opened.

Choosing the Right Camera Film

There are many types of 35 mm film available to use for astrophotography. With the influx of digital cameras some specialized camera films are fast becoming obsolete.

Film Sensitivity

Choosing a film depends upon which type of astronomical object is being photographed. Films speeds typically of the order 400ISO are suitable for most types of astrophotography, as they provide the best sensitivity without compromising on grain quality. This is because bigger grains tend to react faster than finer grains. Moreover, ISO sensitivity is linked to exposures times, e.g. a 400ISO film will capture images in half the time of a 200ISO film. Also, a fast grainy film will not show as much detail as a slow fine grain film, so you have to compromise if you want a balance between exposure time and quality of the image.

Some color films are more sensitive to certain color range of the spectrum. Fuji film tends to be green sensitive, whereas Kodak film is red sensitive and hence, is more suitable in capturing Nebulae. Black and White film is suitable for photographing the

Taking Images

Moon. It can also be used with color filters to produce a true color image when all the filtered images are subsequently combined together.

All films suffer from reciprocity failure. This is where, after a period of time, the film loses its sensitivity and you cannot capture any more light from the image without significantly increasing the exposure time. I.e. the relationship between the exposure time and image intensity is not maintained. Different films suffer from different levels of reciprocity failure. You can obtain details of these levels from most of the major film manufacturers.

Film Format

Camera film can come in slide or print formats. Slide film is useful if you want to show an audience your results. If you use slide film and you don't process the film yourself, then it is best to have the photo processing outlet provide you with them uncut. You can mount the slides yourself rather than have the processing outlet do it for. The outlet might cut some the images in half, ruining them if they are not familiar with mounting slides containing astronomical images. This is because the astronomical images will be as dark as the frame separation, and thus will be almost invisible to a commercial cutter. Processing prints is different to that of slides, see next section.

Processing the Film

If you are taking the film to a photo outlet for processing, then you need to specify that you are providing astronomical images. You should request that they 'print all' regardless of whether the photos appear blank and show nothing on them. On many occasions, individuals have taken their unprocessed film to the photo outlet for processing, only to find that the outlet has either ruined the images, or have not processed them at all. Your only compensation is that you are not charged by the outlet, since all the images appeared 'blank'.

Some films like the Ektachromes can be 'push processed', i.e. the speed of the film is increased through special processing techniques. You can specify how far you want the film push processed. For example a 400ISO push processed 'one stop' takes the speed to 800ISO, 'two stops' to 1600ISO. You should note however, that push processing increases the graininess of the film so some image detail may be lost.

Some Astronomers process the prints themselves. This gives them greater freedom to enhance the image using special darkroom techniques.

Astrophotography with an LXD Telescope

Of course, to take magnified photographs of objects in the night sky a telescope is required.

Although you can use any telescope in the LXD range for astrophotography, the best ones to use are the SNTs due to their 'fast' focal ratios (explained in Chapter 3). These telescopes with fast ratios significantly reduce the amount of exposure time you need to capture the image of a faint object, and thus obtain sharper images.

There are a number of ways a camera can be attached to an LXD telescope in order to take pictures of objects.

- Piggyback
- Prime Focus
- Eyepiece projection
- Afocal projection

When a camera is attached to the focuser you will probably have to re-balance the telescope. It is best to overbalance the RA axis in the direction of the tracking. This reduces the strain on the motors which now have to manage the additional weight of the camera equipment. Refer to Chapter 4 for balancing the telescope.

Piggyback Astrophotography

In piggyback astrophotography, the camera is mounted on top of the OTA via a camera bracket (see Chapter 12, Figure 12.11). The camera bracket is very similar to the head of a camera tripod, in that you can point the camera to almost any position in the sky.

The LXD telescope tracks the stars, so exposures longer than a minute can be carried out without any star trails forming. In fact wide-field regions of the sky such as the Milky Way can be captured with just a camera and a 50 mm lens piggybacked on the telescope. You can use fixed lenses such as a 135 mm, in order to capture extended objects such as the North American Nebula or the Andromeda Galaxy. However, the more you zoom in the more likely the camera will pick up tracking errors and vibrations of the telescope for long exposures. Therefore, you will need to use the telescope as a guide scope in order to reduce the tracking errors as much as possible. I would advise you not to use a Zoom lens, unless you fix the zoom mechanism at a particular focal length. The reason is that if the camera is pointing at an angle high up in the sky the lens might slowly 'zoom' under its own weight resulting in a smearing 'effect' of the stars on the final image. This produces a good effect though!

Capturing an Image. The following steps will show you how to take a photograph with a camera piggybacked on top of a telescope. It is assumed that the Autostar has been set up, and the tracking mode is activated.

1. Point the telescope to a bright star in the sky. Center the reticle eyepiece so that the star is in the central cross hairs.
2. Point camera on the piggyback mount to a suitable region in the sky. You don't have to point the camera in the same direction as the telescope.
3. Open the camera shutter using the cable release.
4. Start the timer using the Autostar Timer facility.
5. Guide the bright star, keeping it in the center of the field of view (See guiding tips).
6. When the time is up the Autostar will sound an alarm and you can release the shutter.

Guiding Tips for Piggyback Astrophotography. The best eyepiece to use to assist in guiding the telescope during long exposures is an illuminated reticle, so the star can be accurately tracked. A high power reticle such as a 9 or 12 mm is suitable for guiding the

Taking Images

Figure 10.1. Wide field image of Cassiopeia Region of the Milky Way taken with an OM1 piggybacked on an LXD75 SC-8.

telescope. The aim of guiding is to place a star in the centre of the field of view where the crosshairs intersect, and keep it in there using the Autostar arrows keys during the long exposure. Setting the speed on the Autostar to a level (speed 1 to speed 3), will move the star without sudden jerky movements of the mount that might get picked up by the camera, producing a blurred and useless image. Figure 10.1 shows a wide field image of the Milky Way taken with a camera piggybacked on an LXD telescope.

Prime Focus

The feature of the SLR camera allows the user to directly view the astronomical object through the camera's viewfinder. The telescope essentially becomes a 'super' telephoto lens, and exposure times are dependent upon the focal length and focal ratio of the telescope (just like with a standard lens). To carry out prime focus photography, the camera lens is removed and the camera body is attached to the focus mount of the telescope via a prime focus adaptor (Figure 10.2). The adaptor consists of a screw thread, which is attached to the camera through a 'T' ring adaptor.

To increase magnification of the prime focus, a Barlow lens is used in-line between the prime focus adaptor and the focusing tube. Alternately a 2× tele-converter can be

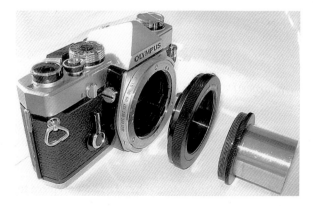

Figure 10.2. Prime focus adaptor and T Ring adaptor.

Figure 10.3. Eyepiece projection adaptor and eyepiece.

attached directly between the camera body and the 'T'Ring' adaptor. Another way to increase magnification is to use eyepiece projection.

Eyepiece Projection

The eyepiece projection method, is primarily used to capture high level detail of the Moon and Planets, due to the long focal lengths involved.

Using Prime focus works quite well for astrophotography, but the magnification and hence, the size of the image depends wholly upon the focal length of the telescope. To obtain higher magnifications, an eyepiece is required to project the image onto the camera plane.

Eyepiece projection is a means of projecting an image from an eyepiece onto the back of the camera plane. Adaptors are available which screw into the camera in a similar fashion to that of the prime focus adaptor. These adaptors are designed to house an eyepiece and place it in the correct position in front of the camera. Of course, you will need to find a suitable eyepiece to fit the barrel of the adaptor, as most eyepieces nowadays tend to be too large in diameter to fit such adaptors. I use a simple Kellner or Orthoscopic eyepiece to fit into the barrel of the adaptor, when I carry out eyepiece projection (Figure 10.3).

Afocal Projection

An alternative to eyepiece projection is using the camera and its own lens to take images. This is placed in close proximity to the eyepiece, so the astronomical image from the eyepiece is projected into the camera. Any type of camera can be used, SLR, digital or video.

This method can be quite tricky as the camera lens has to be exactly aligned with the central optical axis of the eyepiece lens. If not aligned, the image will be 'cropped' and vignetting will occur (objects appear darker near the edge of the field of view). Afocal adaptors are freely available from most astronomical stockists. Figure 10.4 shows a digital camera attached to an LXD 8" SCT using such an adaptor.

The best eyepieces to use for afocal projection are those with long eye relief. Low power two inch eyepieces are ideal for the task, although the size of the objects in the field of view will be quite small. A Barlow lens can also be used to magnify the image. Figure 10.5 shows a shot taken of the Moon using a Nikon Coolpix 5200 digital camera

Taking Images

Figure 10.4. Use of a parfocal adaptor.

placed afocally against a 40 mm Optiplex 2 inch eyepiece. As you can see there is some cropping to the top left of the image, because it was difficult to align the camera with the image in the eyepiece, even using the parfocal adaptor.

Guiding Tips for Prime Focus and Eyepiece Projection

The accuracy of the telescope tracking has to be near-perfect for prime focus and eyepiece projection astrophotography, if you are to produce quality pin-sharp images of objects. I.e. the polar alignment of the telescope has to be precise.

There are several methods which can help you guide the telescope, in order to improve the tracking for prime focus photography. One method is to use an 'Off Axis'

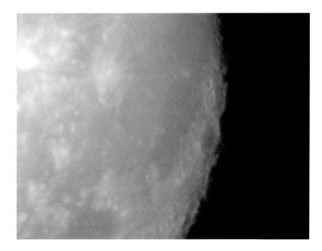

Figure 10.5. The Moon taken through a Nikon Coolpix digital camera.

guider. Another method is to use a small refractor attached piggyback-style in parallel with the OTA.

Digital autoguiding is now generally used by astronomers. The autoguider actively tracks the telescope, so you don't have to continually keep manually adjusting the telescope directional controls during a long exposure.

Taking Pictures – Exposure Settings

For shooting the Moon at prime focus, a typical setup using an SLR camera with 400ISO film will require exposure times from 1/15th to 1/250th second. The exposure time will depend upon the lunar phases.

Photographing the Sun with a solar filter will require a slower film and shorter exposure times typically 1/125th for 200ISO film. It is best to use a slow film in order to capture the granular details on the surface of the Sun, such as Sunspots and Faculae.

Taking pictures of Planets requires more patience if you want to capture minute details on the discs. For example photographing Jupiter using 400ISO film would require exposure times from $\frac{1}{2}$ second to 2 seconds. Longer exposure times will wash out the detail on the planet's disc although some of the satellites orbiting the planet will be recorded.

Experiment with the exposure times by 'bracketing' them. Take a photo at one shutter speed (e.g. 1/125th second) then take photos one either side of the speed (1/60th and 1/250th), one of them is bound to produce a good result.

Digital Imaging

Digital Imaging, of astronomical objects is a discipline that has recently become very popular amongst amateur astronomers. Digital imaging techniques have vastly improved over the past few years. The 'digital darkroom' has replaced the traditional photographic darkroom.

The biggest advantage over traditional methods is that digital imaging gives you *instant* results. Traditional methods are fast becoming obsolete. Soon the day when you take your astro-snaps to the local photo processing outlet, and wait patiently for several hours or days for them to be processed will be long gone. Digital imaging allows you to see your results live, immediately as they are captured. You can discard the poor quality ones and keep the best. You can then digitally enhance the images in the digital darkroom with a personal computer and produce stunning results.

There are many types of digital devices available, which you can attach to the LXD telescope. Some are more dedicated for astro-imaging than others.

- Web Cameras
- Charge Coupled Devices (CCDs)
- Digital SLRs

CCD cameras have shown that even amateur astronomers can take images on par with the professionals. Web cameras are the cheaper alternative to CCDs and can produce images just as spectacular.

Taking Images

Webcam Imaging

The use of Web cameras, is a great way of being introduced to digital imaging. Off-the-shelf webcams are freely available, although for astro-imaging, you need to consider a number of factors when choosing the right webcam.

- Sensitivity to faint objects
- Suitability for connecting to a telescope
- Pixel Resolution

Some webcams are more light-sensitive than others, and the better they respond to low light levels, the more suited they are to astro-imaging, which is of course, recording very faint objects. The most suitable webcams, are those that can respond to light levels around or below 1 Lux. (Lux is a unit measurement of light intensity). Some models of Web cameras can be modified in order to increase their light sensitivity. This can make them almost as good as CCD cameras.

Most Webcam lenses detach from the main body. The lenses are replaced with a special adaptor, which allows the Webcam body to be attached directly to the focusing tube of the LXD telescope.

The majority of Webcams have a resolution of 640 × 480 pixels. These are generally okay for imaging some objects such as the Moon. However, the more pixels the Webcam has, the better it can resolve details on an object.

Of course, manufacturers are bringing out more sophisticated Webcams all the time. Probably the best Webcam 'currently' on the market, most suited for astronomical imaging is the Phillips ToUcam Pro (Figure 10.6).

Meade has produced a dedicated webcam-like device for astronomical imaging called the Lunar Planetary Imager (LPI), shown in Figure 10.7. This device essentially acts like a Webcam, although as the name suggests it is best suited for the Moon and the Planets, and some of the brighter deep sky objects, like the Orion Nebula and the Andromeda Galaxy. The LPI also functions as an autoguider. For this to work, you

Figure 10.6. ToUcam Pro with prime focus adaptor and shown fitted to telescope (inset).

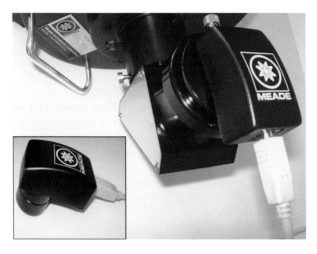

Figure 10.7. Lunar planetary imager.

need to have the Autostar connected to the same PC as the LPI. This is so that the Autostar suite included with the LPI package has access to the telescope's controls.

One thing you need to consider when using Webcams is that the frame size of the CCD chip, which in general is limited in size. So, the field of view that is imaged will be small. Hence, it may take some time and practice to align the object up in the small field of view and keep it there in order to attain an image.

There is an excellent discussion group QCUIAG on the internet dedicated to Webcam astronomy, which provide useful information and tips on using all types of Webcams and digital imaging (see Appendix D for website details). I would also recommend Martin Mobberley's 'Lunar and Planetary Webcam Users Guide'.

CCD Imaging

CCD cameras, such as the Meade Deep Sky Imager, SBIG ST series, SAC and Starlight Express are available to the more serious digital imaging enthusiast. CCD imaging is best for telescopes with fast optics, such as the LXD SNTs. The SCT can be used for CCD work but it requires an f3.3 focal reducer attached between the telescope and the CCD in order to get the best results.

CCDs are a step up from Webcams. They are more sensitive to low light conditions, have CCD chips with greater light collecting areas. They also have the capability to integrate just like a traditional SLR camera. CCD imaging chips tend to be more sensitive in the infra-red region, so requires cooling in order to improve the signal to noise ratio (SNR). IR blocking filters are usually built into the lens or mounted behind it in order to prevent focus problems, and color dilution problems for color CCDs. SNR is also improved by taking a 'dark frame', before capturing an image and then subtracting the 'dark' frame from the image during processing.

Most CCD cameras tend to be black and white. To produce color images, you need to take three independent images using color filters. The images are then combined together, through stacking in order to produce a single color image.

Taking Images

Standard color filters normally used are Red, Green and Blue (RGB). Sometimes RGB are combined with a black and white Luminance (L) image to produce an LRGB image. This is higher in contrast than the standard RGB image. More recently, filters not normally associated with deep sky imaging are being used by CCD enthusiasts such as H Alpha, Oxygen III and H Beta filters. Combined together, these produce true color rendition of the object, more enhanced than the standard RGB images. One-shot RGB color CCD cameras, such as the Meade DSI or the Starlight Express MX-7C produce full color images with just a single exposure. The single shot color cameras however, compare poorly to their high resolution superior black and white CCD counterparts. Even so, they still produce excellent quality images.

To achieve images on par with those that are seen in astronomy magazines, exposures of a length similar to traditional astrophotography methods are normally required. However, with advanced image stacking software such as Registax, you can take multiple short exposure shots and 'stack' them all together as if it was one long exposure. E.g. ten times 60 second exposures are equivalent to one single 10 minute exposure. If you are using color filters to generate a color-combination image, you will need to carry out different time length exposures for each filter, the Red filter being the shortest exposure and the Blue filter the longest.

Many CCDs have the capability to guide the telescope whilst an image is being produced. These CCDs employ a small portion of the total image frame. If a small CCD chip size is used, it can be detrimental to the quality of the final image, as some of the pixels normally used to capture the image are diverted for autoguiding purposes. To combat this problem, some astronomers use two CCD cameras; one dedicated for autoguiding and the other dedicated for imaging.

Volumes of material have been written on the subject of CCD imaging. If you are new to digital imaging I would recommend Adam Stuart's "CCD Astronomy. High-Quality Imaging from the Suburbs" which provides an insight to what you can achieve even with modest imaging equipment.

Digital Cameras

Most people own a digital camera. When they were first introduced, it wasn't long before astronomers started placing them afocally in front of an eyepiece, in order to capture bright objects like the Sun Moon and Planets. Zoom and Macro features on these cameras allow you to obtain close-up images of the object in the eyepiece.

More recently digital SLRs have come on the market. These act much like traditional SLR cameras, but house a large format CCD chip instead of 35 mm film. They are more expensive than their traditional counterparts. The Nikon D70 and Canon EOS 20D digital SLR cameras provide full color RGB images using large 6 or 8 Megapixels CMOS chips. In fact recently, Canon acknowledged the fact that an increasing amount of astronomers are using the EOS 20D for astrophotographic purposes. So, they brought out an enhanced 20Da model that was more suited for astrophotography. Unfortunately, these particular models are no longer commercially available. New and second hand ones are few and far between. Astronomers will have to wait to see if another model will be introduced, dedicated for astrophotography.

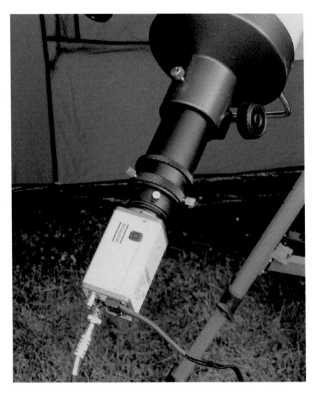

Figure 10.8. Mintron video camera connected to an LXD telescope. Image courtesy of Alan Marriott.

Video Cameras

Webcams not only take still images but can also be used for real time video imaging. Attaching the webcam to the telescope and setting the mode to 'live' video, you can record objects in real time as they are seen through the telescope. However, most webcams do not provide as high quality video as dedicated video equipment.

There are many video cameras, on the market to choose from. CCD video Cameras such as the StellaCam 2, Mintron are suitable for recording objects such as the Moon, Meteors, Satellites and Planets. Some deep sky objects are within their reach as their light sensitivity can be low as 1/2000th of a Lux. Video output is primarily displayed on a TV or VCR but to display the output onto a PC monitor or laptop a video capture device is required.

The video cameras are attached to the telescope using a prime focus adaptor, which is normally supplied with the camera (Figure 10.8).

Frame grabbing software like Registax can capture a single frame from the video recordings. This is especially handy for grabbing the best frames when the seeing conditions are near perfect and the object is pin-sharp during fleeting moments. The best frames can then be added together digitally to produce a high contrast image.

Taking Images

Focusing Tips

Focusing an object through a camera or imaging device can be very difficult at times. All it takes is a slight tweak of the focus in the wrong direction, and you end up with blurred images.

Focusing is probably the most important thing you need to master when undertaking astrophotography. There are many useful focusing aids now available to help you achieve pin sharp images. For the LXD SC-8 a Hartmann Mask can be used. The mask is basically a circular piece of black cardboard with two or three perfect circular holes cut out of it. When placed over the front of the SCT the Hartmann mask produces two or three images of the object, depending upon the number of holes used. As you get closer to the focus, you will see the images merge, until only one sharp image remains. You can purchase Hartmann masks from many astronomical stockists. Alternatively, you can easily make your own. Mike Weasner's Mighty ETX website explains how to construct a home-made Hartmann mask.

The SNT and Refractors use different focusing aids to that of the SCT. Examples are knife-edge testing, flip mirror systems, parfocal rings and PC software focusing tools such as the Magic eye utility (part of the LPI Autostar suite).

Image Processing

Once you have obtained your raw images or video footage you need to process them using digital darkroom techniques. There are many image processing packages available, such as Registax, Photoshop, Maxim DL, IRIS, K3CCD Tools, AstroStack to name but a few. Some are dedicated for astronomical image processing, like Registax and K3CCD Tools. While others like Adobe Photoshop are dedicated professional graphics packages.

Registax is probably one of the most powerful and popular astronomical image processing packages currently available. It has the capability to process raw video footage (AVI format) and automatically stack the best frames, even if the object drifts slightly. Registax has extensive features that would take a whole chapter (or more) to explain. My advice is to download it for free and try it out for yourself (see Appendix D for website address details).

Maxim DL, is another good package. You can carry out image processing, to get the best out of your images, using the package's extensive list of features. It will also control your telescope through an RS232 interface to perform autoguiding, control various models of CCD cameras, focus control and even filter wheel control. Of course, something like this with so many features does not come cheap, with the full package costing over $500 and the image processing package costing around $200. However if you are really serious about image processing, then it is something worth investing in.

Conclusion

There are numerous books as well as internet resources available on the subject of astrophotography and digital imaging (see Appendix D). They provide a vast amount of information on every aspect of imaging, and go into much greater detail that was discussed in this chapter.

With the right imaging equipment, it is possible to take spectacular images through your LXD telescope. Astrophotography is the next logical step from observing, although it is not for everyone. I would not expect beginners to jump in at the deep end and attempt any type of advanced astrophotography without trying out some simple techniques that I have discussed earlier in this chapter.

CHAPTER ELEVEN

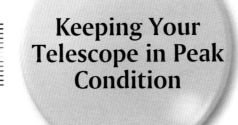

Keeping Your Telescope in Peak Condition

You need to take special care of your telescope and its optics if you want to get the best out of it. This will prolong its operational life and provide you with many years of enjoyment.

There are many good books and internet resources available on cleaning optics and maintaining a telescope. One such book 'Care of Astronomical Telescopes and Accessories' by M Barlow Pepin is well recommended.

In this chapter, I will discuss how to align and clean the optics and talk about basic telescope maintenance and storage for your LXD telescope.

Collimating Your LXD Telescope for Pin-Sharp Images

The quality of an image through a telescope depends upon a number of factors; seeing conditions, quality of the optics and the collimation of the optics. Collimation is the accurate alignment of the optical axes of all the components in the telescope.

With the exception of a few telescope models such as small refractors collimating is something that you cannot avoid during the time you own a telescope. To achieve pin-sharp images the telescope's optics need to be critically aligned with one another.

A telescope can lose its collimation for many reasons. The telescope could have been accidentally jolted, or vibrations could have affected the alignment of the internal optical components during transportation. Even thermal flexure of the telescope tube can affect collimation.

Collimation techniques vary between different LXD telescope models due to their different designs. It is advisable to get an extra helping hand to carry out the collimation where, one person looks through the eyepiece and monitors the effect of the other person carrying out fine adjustments of the optics. Of course you can carry out the collimation on your own but it might take a little longer to complete the task.

There are lots of resources available discussing collimating a telescope from various books, the Internet, and of course the manual supplied with the telescope. These resources provide much more detail than I intend to supply here.

Testing the Collimation of a Telescope

So, how do you determine the alignment of a telescope's optics? Well, you carry out two methods; the use of Collimation tools during the day for rough collimation, and then a Star Test at night for fine tuning the optical alignment.

Star Testing (ECT: 15 Minutes)

Star testing a telescope is a means for determining the collimation of the optics. For first light, a relatively quick star test is carried out just to check that the collimation is okay, before moving on to setting up the telescope for a night's observing.

The best way to test the collimation of a telescope's optics is to use a point-like source such as a star. When a star is defocused through a telescope a circular disc is seen. For telescopes like the LXD SNT, SCT or Reflector, having a central obstruction secondary mirror, a shadow in the shape of the obstruction is seen at the centre of the defocused image. The LXD AR refractor however, has no such central obstruction near its primary lens, so the shadow in the centre of the circular disc is not seen.

Upon closer inspection of the defocused star image, you will notice a set of alternating bright and dark rings increasing in radii from the centre of the image. The shape of this pattern determines the collimation of the optics (Figure 11.1). In fact, you can determine various problems of the optics through investigating the shape of either side the focus of an image.

A point source is used rather than an extended object to perform the test e.g. Moon or Planet. Extended objects do not reveal concentric patterns.

Seeing conditions play havoc with the defocused image of the star so, you should try to perform the star test when the atmosphere is relatively calm, as this can affect the results of the test. This is easier said than done, and you don't normally have the luxury of choosing a calm night for an observing session. Ground surfaces contribute to atmospheric turbulence, so you should try to avoid viewing the star over a hard-standing area, such as concrete patios, buildings or roads.

You should also ensure that the telescope has been cooled down before you start the alignment test. Air currents within the telescope tube causes significant problems with discerning the defocused star image. This cooling could take as long as 30 minutes or so.

Keeping Your Telescope in Peak Condition

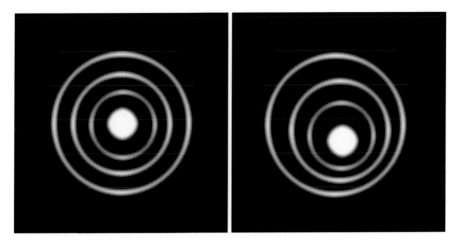

Figure 11.1. Airy discs: Collimated optics (left), uncollimated optics (right). Image courtesy of Alan Marriott.

For perfect collimated images, concentric rings are seen which are symmetrical in shape about the point of focus.

You should have already carried out a rough focusing test of the optics during daylight hours, by focusing on a distant object greater than 500 meters away, such as a television mast or tree. If you found that you had problems focusing the object, then you will need to investigate the problem and collimate the optics.

The steps for carrying out a basic star test are as follows:

1. Point the telescope at a bright star which is at least 50° above the horizon. Centre the star in a 26 mm eyepiece and slightly defocus it until you see the defocused ring pattern. Make sure that the star stays positioned in the central region of the field of view. This is because some eyepieces distort the circular shape of the ring pattern at the edge of the field of view (known as 'coma'), making the optics appear misaligned, where only slight collimation adjustments are actually necessary.
2. Switch to a high power eyepiece and focus the star again. The rule of thumb in determining the magnification used to see the focused ring pattern of a star is about 1× per cm of the primary aperture e.g. for the SC-8 (203 cm) it should be about 200×. Single high power eyepieces e.g. 9 mm, are used to obtain high magnification images, rather than Barlow lenses, which can produce additional irregularities to the focused ring pattern.
3. Note the shape of the ring pattern as you move through the focus. You will find that a focused ring pattern is normally very small unless you use very high magnifications. This means that the telescope will have to be set up to track the sky before any star testing is carried out.

If you are not satisfied with the quality of the image hence, the collimation of the optics, then the optical alignment will require further tweaking.

Collimation Tools

There are several different types of collimating tools to help you attain a near-perfect optical alignment of your telescope. The tools can be used for any of the LXD range of telescopes.

Specialised collimating eyepieces use internal mirrors in order to produce more internal reflections to aid in collimating the optics. The collimation device called a *Cheshire* eyepiece is commonly used to collimate all types of telescopes including refractors. Looking through a Cheshire eyepiece, you will see a number of reflections of the internal primary and secondary optical elements. At the very centre of the view a small circle is seen. When the optics are aligned, this small circle reflecting light will appear to go dark, implying that the optical pathway internal to the telescope is closed. In other words, the mirrors inside the Cheshire eyepiece are reflecting the light path back through the telescope's optics. Hence, the optics will be perfectly aligned with one another.

Another tool becoming popular is laser collimator, although more accurate than traditional collimating eyepieces, it is more expensive. They tend to be more expensive than traditional collimating eyepieces, although more accurate. The laser collimator is attached to the focuser of the telescope using a standard fitting and a thin red laser light is fired through the optical elements. Sensors surrounding the laser gun determine the optical alignment of the telescope.

A rough and ready way of collimating a telescope is by taking a film canister and piercing a small hole through the centre of it. The hole should be big enough to see all of the reflected components inside the OTA, which should all appear concentric.

Collimating the AR Refractor (ECT: 1 to 2 Hours)

The majority of refractors on the market do not need collimating as their primary objective lens is fixed in the OTA. However, large refractors nowadays have the ability to fine tune their optics with the AR model being one of those telescopes. The LXD75 AR-6 and AR-5 models and the LXD55 AR-6 model are able to be collimated, whereas the LXD55 AR-5 model has no such ability.

The AR-6 Refractor is probably the most straightforward instrument to collimate compared with other telescopes in the LXD range, although collimating is not usually carried out as often as the other telescopes.

Remove the dust cap from the front of the OTA and inspect the primary objective lens housing. You will see that the lens cell has three pairs of Phillips head screws located at 120° apart from each other (Figure 11.2). The set of screws, labelled 'A' are loosened to allow the other screws labelled 'B' to adjust the objective lens.

Collimation is best done in daylight hours. The following procedure uses a Cheshire-style eyepiece tool as an example.

1. Remove the Dust cap and the diagonal from the OTA.
2. Use a bright artificial source (*NOT the Sun*!) like a torch or simply point the telescope tube at the bright sky during the day.

Keeping Your Telescope in Peak Condition

Figure 11.2. AR-6 primary lens collimation screws.

3. Insert the Cheshire eyepiece into the focuser. What you should see is a single dark circle at the centre of the field of view. If there appears to be two circles either separated by a very small distance or overlapping (usually the case) then adjust the Phillips head screws at the front of the tube to bring them to a single image. It's as simple as that!

Collimating the SNT (ECT: 1 to 2 Hours)

Short focal length SNTs tends to be more involved compared to SCTs or Refractors, when carrying out collimation of the optics. This is because there are two sets of collimation screws to adjust, one set for primary mirror and the other set on the front corrector plate for the secondary mirror (Figure 11.3). There is also a small circle located at the centre of the primary mirror to aid with laser collimation. The circle was placed on the mirror during factory assembly.

1. Orient the OTA so that it is horizontal with the ground. Make sure that the focuser is in a comfortable viewing position and that the primary mirror is located to the right of your position.
2. Look through the focuser with the eyepiece removed. You will see reflections of the OTA internal components (Figure 11.4). Note that the secondary mirror position is deliberately offset away from the focuser and towards the primary mirror. This offset is the normal configuration for short focal length instruments like the SNT.
3. Remove the front dust cap (Figure 11.3, lower right) covering the collimation screws located in the centre of the front corrector plate.

Figure 11.3. SNT collimation screws (front and back).
Image courtesy of Chris Williams.

4. Carry out the collimation. Table 11.1 provides a summary of collimating screws you need to adjust in order to align the relevant components.

You may not be able to choose the correct screw to adjust the secondary or primary component on your first attempt. Some trial and error is called for before you get the feel of which screw to adjust.

Normally, when you loosen or tighten one of the screws, you have to subsequently adjust the other two screws to ensure that the secondary or primary mirrors are still secure in their housings. This also aids with the tilting of the mirror by allowing it to move more freely. In the case of loosening the secondary diagonal screws, ensure that

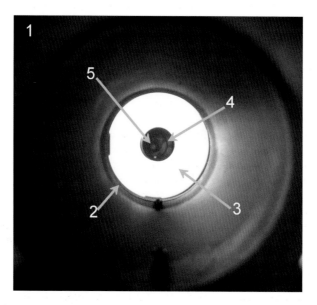

Figure 11.4. Internal elements of the OTA seen through the focuser.
Image courtesy of Chris Williams.

Keeping Your Telescope in Peak Condition

Table 11.1. SNT Collimation Description Summary

ID	Description	Set of Screws to Adjust
1	Focus Drawtube.	None
2	Diagonal mirror: The mirror is not centrally aligned with the focus drawtube.	Adjust S1 screws of the secondary diagonal mirror holder on the front corrector plate.
3	Reflection of primary mirror: The reflection is not centrally aligned with the secondary diagonal mirror (2).	Adjust S1 screws of the secondary diagonal mirror holder on the front corrector plate.
4	Reflection of secondary diagonal mirror: The reflection is not centrally aligned with the primary mirror reflection (3).	Adjust P1 screws on back of primary mirror. Note: There are an additional three screws on the primary mirror housing which lock the mirror in place (labelled P2 in Figure 11.3). These screws should be loosened before using the main collimation screws to adjust the tilt of the primary mirror.
5	Central Collimation Circle: The spot is not in the centre of the secondary diagonal mirror reflection (4).	Adjust P1 screws on back of primary mirror.

you do not unscrew the screws too far out so that the secondary diagonal becomes loose in its holder.

You need to take special care when looking through the focuser, at the same time you are adjusting the secondary diagonal mirror screws on the front corrector plate. You may accidentally mark or scratch the surface of the plate with a tool or touch it with your fingers. Unfortunately, this happens quite often.

Collimating the SCT (ECT: 1 to 2 Hours)

Unlike the SNT, the SCT has a single set of collimation adjustment screws located on the front corrector plate (Figure 11.5).

From Figure 11.5 you can see that adjusting the collimation screws could be tricky with just one person. Special care has to be taken to prevent the front corrector plate being marked by an Allen key, especially when you are carrying out the collimation on your own in the dark. One solution is to replace all three screws with a set of collimation knobs, so that you can adjust the secondary using your fingers instead rather than an Allen key.

A Cheshire eyepiece can be used to collimate an SCT, however to produce more accurate collimation, you should carry out the alignment process with a point like source such as a star.

To collimate an SCT perform the following steps:

1. Set up the telescope as normal. Set the Autostar to simple astronomical tracking mode with or without the Goto setup.
2. Remove the star diagonal from the focuser.

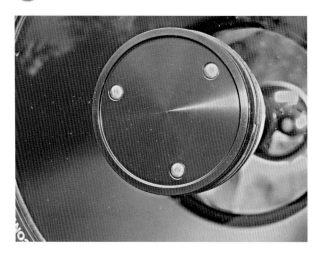

Figure 11.5. SCT secondary mirror collimation screws.

3. Point the telescope to a star which has the least amount of motion over a long period such as Polaris or in the Southern hemisphere a bright star near the south celestial pole.
4. Centre the star in an eyepiece (26 mm) and perform a star test by slightly defocusing the image. If the seeing condition is good then you should see whether the set of rings are concentric or not. If the rings are not concentric and the central black spot is to one edge of the image, then you need to determine which of the three hexagonal screws you need to adjust first.
5. Of course, you could simply guess which screw you want to adjust, but that could lead to further misalignment and take longer to collimate. To determine which screw to collimate, the following method is carried out.
   - Place a finger on one of the collimation screws at the front of the OTA. When you look through the eyepiece you should see a blurred shadow of your finger.
   - Carefully circle your finger around the secondary housing until the shadow of your finger you see in the eyepiece coincides with the black spot located to the edge of the image. Ensure you don't accidentally touch the surface of the corrector plate.
   - The screw that is the closest to your finger is the one to be adjusted first. If your finger is in between two of the screws then you should adjust each one.
6. Make fine turns of the collimation screw to try and bring the black spot to the centre of the rings in the image. You will notice that the image will shift in the field of view, as you make the adjustments and may move out of the field of view altogether, if you turn the adjustment screw too far. You must re-centre the image in the field of view after each successive turn.
7. Loosening a screw will require tightening of the other two screws in order to continue to secure the secondary mirror to the front plate.
8. Continue until the image is a perfect set of concentric rings. Try a higher power eyepiece and repeat the Star Test to check if the collimation is precise. Repeat above steps where necessary.

Keeping Your Telescope in Peak Condition

Figure 11.6. N-6 reflector collimation screws. Image courtesy of Kevin Downing.

Collimating the N-6 Reflector

Collimating the LXD reflector is similar to that of collimating an SNT. There are two set of collimation screws for the primary and secondary mirror housings (Figure 11.6). The lack of a corrector plate implies you can place your hand inside the OTA and adjust the secondary mirror.

Collimation of the Reflector is best done during the day which you can then switch to star testing at night to fine tune the optics. Once again like the other telescopes, if you tighten one of the screws either on the primary or secondary mirror housings, you will have to loosen the other two in order to provide enough travel distance to adjust the mirror.

The steps are as follows:

1. Orientate the OTA horizontally so that the primary mirror is to your right.
2. When you look through the focuser with you will see the internal components of the Reflector (Figure 11.7).
3. Adjust the collimation screws on the primary and secondary housings accordingly. (See Table 11.2.) There are three mirror lock thumbscrews on the back of the primary mirror housing, like the SNT. These must be loosened before using the collimation screws.
4. All of the reflected elements should be concentric to produce a near-perfect collimation.

Collimation Summary

Collimation requires some experience in interpreting the images that you see in a defocused star. In fact, you may need to re-align your finderscope after you have performed the collimation.

Special care must be taken when adjusting the optics of the telescope as you can easily damage the components if not handled correctly. Always use the correct tools

Table 11.2. N-6 Reflector Collimation Description Summary

ID	Description	Set of Screws to Adjust
1	Focus drawtube: Does not require adjusting.	None.
2	Reflection of the primary mirror: The reflection is not completely seen within the area of the secondary mirror (3).	Loosen all S1 screws and gently twist with one hand the entire secondary until the reflection of the primary mirror (3) is centralized with the secondary mirror (2). Tighten S1 screws when done.
3	Secondary Mirror and Holder[a]: Position is Left or Right of the centre of the drawtube (1). Position is Above or Below the centre of the drawtube (1).	
4	Secondary Mirror Vanes.[a]	
5	Observers Eye: Position is not centre to the reflection of the primary mirror (2). (The central area is a small disc if using a Cheshire collimation-style eyepiece). Note: Figure 11.7 shows a camera lens reflection, which took the original image.	Loosen the three mirror lock screws P2. Adjust primary mirror tilt screws P1 to align eye/spot to centre of primary mirror reflection image. Tighten P2 screws when done.

[a] The LXD75 manual states you can adjust the vanes in order to align the secondary mirror holder with the centre of the focus drawtube. This is in fact not the case. The front OTA ring and vane assembly is actually a single cast unit, and hence not adjustable. If the secondary mirror is significantly misaligned with the drawtube, then you will need to return it to Meade for repair.

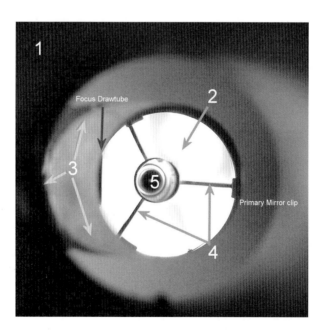

Figure 11.7. Internal elements of N-6 reflector through focuser. Image courtesy of Kevin Downing.

Keeping Your Telescope in Peak Condition

and never over-tighten the screw/bolt heads, as replacements may be difficult to obtain.

Cleaning the Telescope

There will come a time when you will feel that the telescope is in need of some 'spring cleaning'.

Dust tends to gravitate toward telescopes if not used for a period of time. An anti-static (yellow) duster is best used to gently wipe over the LXD mount and tripod exterior surface. The LXD75 tripod shiny legs can be polished if so desired with a gentle cleaning polish found in your local superstore; although over polishing may remove the manufacturers plating!

Cleaning the Optics

Over a period of time the optics will attract dust and moisture marks although this will not normally interfere with the quality of the image.

Cleaning the optics is usually done once a year, perhaps longer. Of course cleaning 'emergencies' will crop up from time to time e.g. accidental liquid spillage or other substances, and you will have to act immediately to save your optics. Ensuring that the optics are covered with dust covers and caps when not in use will prolong the need for not cleaning.

The following do's and don'ts provides some guidance on cleaning your telescope, optics and accessories. Many telescope distributors now sell telescope optic cleaning kits. However, you should still err on the side of caution when cleaning delicate surfaces, with any type of advertised lens cleaning solution.

Cleaning Objective Lens and Corrector Plates

LXD Models: AR, SNT, SCT.

1. *Do* use a camera blower bulb and lens brush to blow the dust off (not your wet breath!). This should be done whilst the telescope is tilted slightly towards the ground, to prevent the disturbed dust from accumulating back onto the lens.
2. *Don't* use compressed air to blow the dust. Some cans contain corrosive materials, which could affect the special coatings on the lens or corrector plate.
3. *Don't*: If the dust is stubbornly adhered to the lens, don't wipe the lens with a cloth (even a lens cloth), as the dust will act as an abrasive and may scratch the lens.
4. *Do* use a mixture of Isopropyl Alcohol (3 parts) and Distilled water (1 part) solution to gently remove finger prints and stubborn marks. Add a drop of washing up liquid (one drop per litre or so of solution). Use cotton balls or soft non-perfumed plain tissue wipes with the solution to gently stroke the lens in a single wipe action across the lens, starting from the centre and working your way out to the edge. Discard each tissue after a single wipe. You will use several tissues before the

entire lens has been cleaned. Allow the lens to dry naturally (e.g. do not use a hair dryer).
5. *Don't* be tempted to take the front optics out of its housing to clean even if you think that the surface on the inside of the tube is dusty.

Cleaning Mirrors

LXD Model: N-6 Only.

1. *Don't* attempt to clean the primary or secondary mirrors of the SNT and SCT models by removing them from the telescope tube, you will struggle to fit them back, as these are fitted and aligned during factory assembly. There should be no reason to clean these mirrors as there will be little dust inside the sealed telescope tube.
2. *Do* clean the Reflector Mirror occasionally, as there will be a build up of dust and deposits since the telescope is not a sealed unit unlike the SNT and SCT models.
    a. Remove the back mirror housing from the OTA. It is best to also remove the mirror from the mirror cell.
    b. Fill a sink with warm (not hot) water. Add two drops of washing up liquid and mix until there is a good amount of soap suds.
    c. Set the mirror into the sink the sink, reflected coating face up. Allow to soak for 2 to 3 minutes.
    d. Remove the mirror from the sink and check the surface for any residual marks. If there are still marks, place the mirror back into the sink and very gently drag a cotton ball across the mirror. Only use the cotton ball once. If there are more marks, use a clean cotton ball.
    e. Remove the mirror from the sink and tilt to drain off the excess water. Wash the sink out and pour distilled water across the mirror surface to rinse any soap suds which could cause marks when dried out. The distilled water rinse needs to be done immediately, as if it is left to drain first there is an opportunity for residue to dry on the surface before the rinse.
    f. Place the mirror on a flat surface, preferably on a soft cloth to absorb runoff and allow it to dry naturally.
    g. Place the mirror back into the mirror housing and re-attach it to the rear of the OTA.

The secondary mirror of the reflector is not cleaned very often. If there is a substantial amount of dust on the mirror then point the telescope towards the ground and gently blow the dust off using a camera blow bulb. Wait a few minutes until you are sure that the dust will not travel down the tube and settle on the primary mirror, when you point the tube upwards again.

Eyepiece Cleaning

Eyepieces require cleaning more frequently than telescope optics as they are handled more intimately than the objective lens and corrector plates. Their glass surfaces tend to be covered in fingerprint marks and grease from eyelashes which will eventually corrode the coatings, if not cleaned regularly.

Keeping Your Telescope in Peak Condition

A similar solution (isopropyl alcohol) is used to that of cleaning the objective lens and corrector plates. Dip a cotton bud into the solution and gently wipe the eyepiece in a circular motion, starting from the centre and working out to the edges. Take care that there is not too much solution on the cotton bud, as the liquid could work itself between lens elements that make up the eyepiece. Turn the eyepiece upside down and repeat the same cleaning action for the lens element inside the eyepiece barrel. Make sure you don't touch the inside of the eyepiece barrel, as you might remove some of the black internal barrel coating and accidentally put it on the lens surface. Always wash your hands thoroughly beforehand (preferably with a degreaser), since otherwise more dirt and body oils may be deposited than removed.

A good test to check that the surface of the eyepiece is clean and free of marks is to very gently breathe on the eyepiece to fog the surface. Any marks will show up and can be cleaned accordingly. On a clean eyepiece this fog should evaporate immediately.

A useful tool for cleaning eyepieces is a LensPen (see website address in Appendix D). The LensPen has a brush at one end and a special non-scratch tip at the other. Use the brush first to remove the dust, then use the non-scratch tip to clean the lens surface.

General Maintenance

Many astronomers like to tinker with their telescope when they are not using it to observe with. Carrying out periodic maintenance prolongs the operational life of the telescope and helps prevent problems occurring.

If the telescope has been used regularly then components such as screws, nuts and bolts tend to loosen and parts wear if they are not looked after properly. Conversely, a telescope that has been sitting in a garage for many months, it will also require maintenance in order to bring it back to peak operational performance.

Motors

- Adjust RA and Dec Motor alignment – See Chapter 4.
- Adjust RA and Dec Cog within Motor Housing – See Chapter 4.

OTA

- Adjust focuser parts to ensure smooth operation – See Chapter 4.

Power

- Check the power status of battery cells of the polar alignment viewfinder.
- Check the power status of battery cells that supply the control panel of main telescope. (Use the battery status utility on the Autostar – see Chapter 13).

RA Axis Movement

- Check the friction between the Dec section of the mount and the RA section of the mount. Remove two rubber tips that cover the holes, which contain two hexagonal bolts (Figure 11.8 left image arrowed. Note: only one bolt is shown as the other one is on the opposite side of the RA shaft). Use a long Allen key and adjust the bolts to achieve the appropriate level of resistance, when the RA

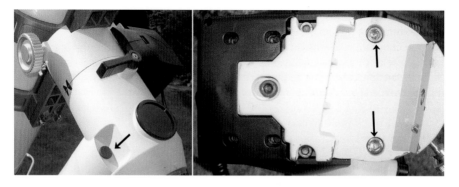

Figure 11.8. RA (left) and Dec (right) axes friction adjusting bolts.

lock is loosened and the RA axis is rotated. Remember to replace the rubber tips back into the holes when finished.

Dec Axis Movement

- Check the friction between the Dec axis cradle slot and the Dec section of the mount. Adjust hexagonal bolts using an Allen key (Figure 11.8 right image arrowed), to achieve the appropriate level of resistance when the Dec lock is loosened and the Dec axis is rotated.

The Ultimate Telescope Tune-up!

This section cannot be fully complete without mentioning Hypertuning or Supercharging the LXD55 and LXD75 telescope. This involves completely stripping the entire mount down to its constituent components, then rigorously cleaning and polishing the gear wheels and the RA/DEC drives as well as other parts of the mount. The components are then re-greased with superior lubricants and then re-assembled for optimum performance.

You can carry out Hypertuning or Supercharging the LXD telescope yourself. Alternatively, if you are not that mechanically inclined or confident that you can put all of the components back together again, you can send the telescope away, to get it done professionally. Both services are available in the United States through Richard Harris's LXD55.com (hypertuning) and Dr P Clay Sherrod's Arkansas Sky Observatory (Supercharging), see website addresses in Appendix D. Feedback from LXD owners who took advantage of either service found that the performance of their telescopes improved significantly.

Storing the Telescope

You need to find an ideal location where (a) it doesn't deteriorate when it is not in use for long periods, (b) it takes the least amount of time to set up from where it is stored

Keeping Your Telescope in Peak Condition

and (c) security – telescopes are expensive and can be prone to stealing. These factors decide whether the telescope is stored indoors or outdoors.

For many astronomers, it is a lifetime ambition to own an observatory to accommodate their telescope and associated astronomical equipment. However, if the finances are not available to buy one or have the know how to build one, then an alternative place to store it is needed.

Indoor Storage

Whether you live in a house or on the fifteenth floor of an apartment block, there are some basic guidelines to follow for storing a telescope indoors.

- Ensure that all dust caps are firmly in place on the front objective or corrector plate as well as the finderscope and focusing tube.
- The Autostar should be disconnected from the control panel and its lead detached from the handset. This is to prevent connection problems with the Autostar, as a result of constant pressure being applied to the socket on the Autostar or if it gets accidentally pulled out of the sockets. (See Chapter 13 – Troubleshooting and FAQ).
- If using battery cells disconnect the cells from the control panel on the mount. Keep the power cells in a cool dry place away from heat sources.
- Find a location indoors that will not be in harms way so, for example no-one trips up over the protruding tripod legs. (You don't want the telescope to get damaged let alone causing injury to the individual who tripped up over it!)
- Use an anti-static telescope bespoke dust cover (not a cloth sheet) to keep household dust and other material such as pet hair settling on the mount and tripod.
- Store the telescope either as a whole or in separate manageable parts. This depends upon:
    a. If stored in one piece do you have the strength to pick it up and carry it out to your place of observing? These telescopes can be quite heavy when fully assembled so lifting them in one is not recommended. If you cannot carry it out in one piece, dismantling it inside and then reassembling the telescope outdoors, could waste precious observing time.
    b. The amount of storage space you have indoors, so instead of a large and bulky telescope taking up a large amount of space in the corner of the room, each separate part is broken down into their main constituents (OTA, Mount, Counterweights, Tripod legs) and stored separately to save space.

Storing the Telescope Outdoors

Some astronomers store their telescope in a garage or shed. Of course with a simple garden-type shed, you will still have to haul the telescope outside every time you want to do some observing, even if the observing location is only a few feet away.

Figure 11.9.
The astroshed.

An option is to purchase or build an observatory to shelter the telescope in a permanent location. There are a number of advantages to building a fixed location for a telescope:

- Time is not wasted dismantling the telescope indoors then setting it up again in the garden, only to find that it has clouded over!
- It is quicker for the telescope to reach optimal observing temperature. A well-ventilated observatory will keep the telescope at the same ambient temperature as its surroundings.
- The telescope will require polar aligning only once and thereafter requires little or no adjusting to keep Polaris aligned. Unless, the telescope has to be repositioned during maintenance.

Some time ago I decided to build an observatory. Various designs were considered including those of domed construction, but it was deemed to difficult and expensive to build. In the end, I decided to build a run-off roof observatory called the 'Astroshed' (Figure 11.9). The Roof rolls off onto side rails and the sides fold down to allow more area of the sky to be seen. Every two years the Observatory is repainted. Where there is wood damage, it is treated and filled. The wheels in the roof are periodically greased, as well as the lock on the door. Other than that, little maintenance is required. A sequence of images and description of my observatory is on my website (see website list in Appendix D).

There are numerous designs of astronomical observatories; each design having its own advantages and to discuss them is a book in itself. The book 'More Small Astronomical Observatories' by Sir Patrick Moore is highly recommended if you are considering purchasing or building an observatory.

Summary

As you can see there is a lot you can do to keep the telescope in a healthy operational state. Of course taking the telescope apart to perform maintenance may invalidate its warranty. You should check the terms and conditions of the warranty beforehand.

CHAPTER TWELVE

Gadgets and Gizmos

Introduction

No telescope is complete without accessories. There are many available for the LXD series of telescopes and in fact, not just produced by the original manufacturer (i.e. Meade). Many companies have taken it upon themselves to manufacture accessories that enhance the LXD telescope hardware.

This chapter provides a general guide to accessories available for your LXD telescope. The list is by no means exhaustive, and if you want to know more information about buying accessories for telescopes 'Star Ware' by Phil Harrington is a book well worth reading.

Eyepieces

Purchasing an eyepiece is just as important as buying a telescope. The best quality eyepieces tend to be very expensive, some costing more than a quarter of the initial outlay of the telescope.

The quality of an image does not solely depend upon the quality of the telescope optics, but also on the quality and type of eyepiece used. Eyepieces come in various types and sizes, ranging from simple designs with a few internal elements to those with a complex array of elements. The magnification and exit pupil properties depend upon focal length and design type. These are discussed in Chapter 7.

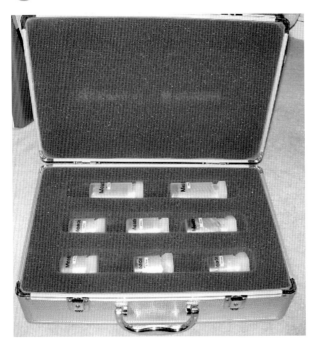

Figure 12.1. Meade anniversary eyepiece kit.

One of the most popular types of eyepieces used by astronomers is the Plössl. It consists of four elements, and is a good all round general purpose eyepiece. The 26 mm eyepiece supplied with every LXD telescope is of Plössl design. A short time ago Meade produced a set of Plössl eyepieces to celebrate their 50th Anniversary. The anniversary kit contains the complete set of Meade Series 4000 Plössl eyepieces ranging from the low power wide field 40 mm eyepiece to the high power 6.4 mm eyepiece (Figure 12.1). The 4000 series has been recently superseded by the superior 5000 series.

Other common types of eyepieces include Kellner, Orthoscopic, Nagler and Zoom. The *Kellner* eyepiece consists of three elements, has good eye relief and is best used at low to moderate magnifications. They have typical focal lengths of between 40 to 20 mm.

The *Orthoscopic* eyepiece consists of four elements and provides excellent sharpness and colour contrast, so are ideal for lunar and planetary work but has a smaller field of view compared to other designs. A disadvantage of orthoscopic eyepieces is that focal lengths 8 mm or smaller, provide pinhole eye relief which hinders viewing of the image. The Orthoscopic eyepiece design has been superseded with superior, more complex designs which offer wide-field and excellent eye relief for relatively short focal lengths. An example is the Televue Radian eyepiece (Figure 12.2).

Zoom eyepieces combine several focal lengths into a single eyepiece design. This is like using several eyepieces at once, although images tend to be slightly dimmer than fixed focal length eyepieces, due to the large amount of optical elements attenuating the light from the image.

Gadgets and Gizmos

Figure 12.2. Televue radian.

Erecting (Terrestrial) Eyepieces and Prisms

The image that is seen in all models of the LXD telescope is inverted. To carry out terrestrial observations you need to see objects the right way up. This is what an erecting eyepiece does. The erecting prism of an erecting eyepiece is similar to a star diagonal (see later in this chapter) except, that it also flips the image the right way up for terrestrial use. The eyepiece magnifies the image by 1.5×.

Barlow Lens

A Barlow lens when placed between the focus tube and the eyepiece effectively increases the focal length of the telescope, hence magnification (conversely, it decreases the focal length of the eyepiece).

The increase in focal length is determined by the degree of divergence (negative) lens used in the Barlow. This in effect, is a multiple of the original magnification. Hence, most Barlow lenses are based on the multiplication factor of the negative lens

Figure 12.3. Two inch Tele Vue 4× Powermate. Image courtesy of Tele Vue.

e.g. 2×, 3×. Barlow lenses of 4× are commonly used nowadays for high magnification digital imaging (Figure 12.3).

A 26 mm eyepiece used with a 2× Barlow lens gives an effective focal length equivalent to a 13 mm eyepiece, but still has the same optical eye relief as that of the original 26 mm eyepiece. Hence, a Barlow Lens essentially doubles the amount of eyepieces you own.

Focal Reducer

As the heading suggests, a focal reducer reduces a telescope's effective focal length, by an amount determined by the degree of convergence. The focal reducer functions in the opposite way to a Barlow Lens, i.e. the magnification is reduced and the overall field of view is widened.

Focal reducers are best used for the LXD SC-8 SCTs, which have a high focal ratio f/10 compared to the SNT telescopes which are much lower.

Two reducers are available f/6.3 (0.63×) and f/3.3 (0.33×). The f6.3 is primarily used for observational purposes, whilst the f/3.3 is dedicated for photographic and CCD imaging, since the exposure time is shorter as a direct result of the focal length being reduced. The f/6.3 reducer can also be used on the LXD AR f/8 and f/9 refractors, but unfortunately not with the SNT models or the N-6 Reflector, as the focusing tube cannot be racked in far enough to achieve a focus.

Star Diagonals

The LXD AR refractor model is supplied with a star diagonal. However, the original diagonals supplied with the LXD55 AR models were of cheap plastic construction, and poorly under-achieved the telescope optics whenever they were used. The LXD75 is supplied with a better constructed diagonal but it is only 1.25 inch in diameter. To get the best results, a 2 inch star diagonal is recommended.

The best quality models currently available are dielectric non-metallic coated mirror diagonals by William Optics (Figure 12.4) or the Televue Everbrite to name a few.

Gadgets and Gizmos

Figure 12.4. William optics diagonal. Image courtesy of William Optics.

Prismatic diagonals tend to produce images less in quality, compared to their mirror counterparts.

Finderscopes

The LXD finderscope is useful for most astronomical purposes. Other types of finderscopes can also be used.

Nowadays many astronomers tend to use 'finders' with zero magnification, rather than traditional 'finderscopes' that use objective lenses. This is because the role of the finderscope, as an instrument used in conjunction with the telescope has changed since the advent of the Goto telescope. In fact, nowadays finderscopes are primarily used to sight alignment stars when an Autostar Goto alignment is carried out.

Finders with zero magnification consist of a reference marker, such as a red laser dot, or red light bull's-eye that is projected onto a transparent glass screen. When peering through the finder it appears that the red dot or bull's-eye is projected on to the sky.

Two common finders are the Telrad (bulls-eye) (Figure 12.5) and the EZ Finder II (Red dot) (Figure 12.6). Both are supplied with bases for attaching to the telescope OTA, which is handy since the LXD telescope does not cater for an additional finder.

Filters

Most telescopes with good optics normally provide excellent images of the Moon, Planets and deep sky objects. In fact, the images can be enhanced even further using special filters. The filters screw into the rear of most eyepieces and are available in

Figure 12.5. A Telrad attached to an AR-6.

$1\frac{1}{4}$ inch and 2 inch diameters. For specialized observational and photographic purposes more dedicated types of filters are used.

Color Filters

Colour filters are used to highlight surface markings or cloud bands on a Planet's disc.

If you are new to using filters then you should start off with a set of basic filters. A typical set comprises of four filters; red, blue, green and yellow (Figure 12.7). They are ideal for first time use. A description of filters to use when observing the Planets is in Chapter 8.

Figure 12.6. Red Dot finder attached to an SC-8.

Gadgets and Gizmos

Figure 12.7. Color filter set.

Color filters are also used to produce true color RGB images from black and white digital imaging devices, such as CCDs.

Light Pollution Filters

Unfortunately we live in an age when the sky is awash with man-made light. LPR filters are designed to reduce or eliminate the sky glow. This is done by blocking high and low pressure mercury and sodium vapour lights, while still allowing light from the rest of the visible spectrum through. Hence, the intensity of the background sky is reduced, so the object can be seen more clearly.

The IDAS filter from Hutech and the Deep Sky filter from Lumicon are two such LPR filters commonly used by astronomers in suburban light polluted skies.

LPR filters are best suited for astrophotography where they increase the length of exposure time before the image 'fogs' over.

Color Correction Filters

These filters are designed to eliminate or at least, reduce the color fringing suffered by short focal length AR refracting telescopes.

Three such filters are the Minus Violet filter from Lumicon or Sirius Optics, Contrast Booster from Baader and V Block Filter from Orion. These filters reduce color fringing that surrounds bright distinct objects, improve the contrast of an image and provide better color rendition.

Polarization Filters

These filters are ideal for reducing the glare of bright objects such as the Moon or a Planet. Variable polarising filters are more commonly used as they can be manually adjusted to suit individual preference (Figure 12.8).

Figure 12.8. Meade #905. Variable polarizing filter. Image courtesy of Meade.

Nebula Filters

These filters utilize particular lines of the visible spectrum emitted by planetary and emission nebulae. They also darken the background sky to make objects more visible.

Three such filters commonly used are Oxygen III, Ultra High Contrast and Hydrogen Beta filters. These three filters are also used for digital imaging to produce a true color image of an object from black and white image capturing devices (see Chapter 10).

The *Oxygen III* (OIII) filter utilizes the doubly ionised Oxygen lines emitted by planetary and faint diffuse nebulae. The filter is used to enhance details of deep sky objects such as the Orion nebula (M42), the Veil nebula and the Ring Nebula (M57) to name but a few.

The *Hydrogen Beta* filter, as the name suggests, utilizes the HBeta line emitted by nebulae such as the Horsehead Nebula (IC 446) and the California Nebula (NGC 1499).

The *Ultra High Contrast* (UHC) filter utilizes both the OIII and HBeta lines which are emitted by emission and planetary nebulae. So it is more widely used than the other two filters. The filter is used to enhance details of objects such as the Orion nebula (M42) and the Dumbbell nebula (M27) amongst many others. The filter is most effective under non-light polluted dark skies.

Filter Wheels

This useful accessory allows multiple filters to be used at the eyepiece. The wheel fits between the focusing tube and the eyepiece and eliminates the need for continually unscrewing the filter from the eyepiece, each time a different one is required.

A filter is changed by simply rotating the wheel. This is an ideal way for comparing views of an object seen through different filters. Some filter wheels are motorised and are commonly used for digital imaging.

Gadgets and Gizmos

Binocular Viewers

Two eyes are better than one especially when resolving details on the Moon or a Planet. This is why binoculars provide great views of the night sky. However, all but the largest binoculars do not have the magnification that telescopes enjoy so their use is confined primarily to just a few astronomical objects and wide-field low power views of the sky. Hence, binocular viewers (bino viewers) are becoming widely popular amongst amateur astronomers.

The Denkmeier Binoviewer caters for most telescope design, including the LXD models. A $1\frac{1}{4}$ inch adaptor is required however, in order for the binoviewer to obtain a focus when using low or high power eyepieces.

Tripods

The LXD55 tripod is unfortunately not as well constructed compared to the LXD75 tripod which is manufactured with stronger materials and is superior at handling the weight of any telescope in the LXD series.

Some LXD owners have gone to great lengths to mechanically reinforce the LXD55 tripod in order to improve stability. Nowadays, purpose-built tripods for the LXD series are commercially available.

The wooden telescope tripod from Berlebach caters for both long and short focal lengths of the AR refractors, SNTs and SCT. Support chains connecting each leg of the tripod increases the overall stability. A useful feature of this particular tripod is that there are markings set at 10 cm intervals on each leg, in order to aid leveling.

A set of Vibration Suppressions Pads from Celestron placed under each leg of the tripod, makes its less susceptible to external vibrations. This results in a dramatic reduction in image movement in the eyepiece, hence a noticeable improvement in image quality.

Motorized Focusers

Critical focusing is important when trying to obtain pin-sharp images of objects for astrophotography.

Adjusting the focuser introduces unnecessary vibrations making it difficult to obtain a sharp focus. Motorized Focusers such as those from Jims Mobile (JMI) retro-fit to the focusers of LXD telescopes, and provide vibration-free, smooth focus of objects (Figure 12.9).

GPS Add-Ons

Site location information is retained when the Autostar is powered off however the date and time settings are lost and has to be set each time the Autostar is powered

Figure 12.9. JMI motor focuser for LXD SNT and N-6 Models. Image courtesy of Jims Mobile.

on. If any of these three parameters are incorrect then the Goto facility will not be accurate.

A Global Positioning System device for the Autostar automatically sets the date, time and location settings received from a GPS Satellite. The GPS is used as part of the Autostar startup process.

Two such GPS add-ons available are the StarGPS from PixSoft Inc. and the GPS Align Mate from Telescope House UK. See Appendix D for website addresses.

Bluetooth Connectivity

The standard method for connecting the Autostar to a computer device is through an RS232 serial cable (see Chapter 9).

With the advent of wireless technology, it was only a matter of time before wireless devices were applied to telescopes. It is now possible to wireless the Autostar RS232 connection through the use of Bluetooth adaptors.

BlueStar from Starry Night is a Bluetooth adaptor that is available (See Appendix D for website addresses). The BlueStar adapter also doubles as a USB adapter, hence removing the need for a USB to Serial cable.

Power Supplies

There are various methods for supplying power to the control panel of the LXD telescope.

Standard D cell batteries drain quickly, if the telescope is slewed often during an observing session. Therefore, you need a good power supply if the telescope is going to last all night.

Gadgets and Gizmos

Figure 12.10.
Extended dew shield on SC-8.

A portable power station contains a heavy duty 12 V rechargeable battery with at least 10 Ah capacity. This generally last a few days or a single night if supplying additional power-hungry accessories.

Of course, where possible an electrical power supply is used in preference to power the telescope electronics. 12 V AC regulated power supplies with at least 1.2 amps ratings are required to power the telescope's electrics. Power supplies are freely available from many electrical outlets such as Radio Shack and most telescope stockists.

Dew Shields and Heaters

All telescopes with a front lens or corrector plate suffer from dew. Longer dew shields and electronic heater systems try to keep optics dew free.

Telescope House UK manufactures extended dew shields for LXD telescopes (Figure 12.10).

Electronic dew heating devices supplied from Kendrick USA or Telescope House UK (Figure 12.11) consist of a heated fabric strap which fits around the diameter of the front lens or corrector plate. An electronic controller maintains the temperature of the strap. These dew systems are very successful at keeping the optics dew free. However, they require a lot of power to maintain a consistent temperature of the optics. Many astronomers prefer to use a separate power supply for them.

Meade #909 Accessory Port Module (APM)

The AUX port on the LXD mount's control panel is something that has generated many discussions amongst owners of the telescope. No-one could find any standard accessories that would directly connect to it, such as auto-guiders or motor focusers. Most owners simply have left it alone.

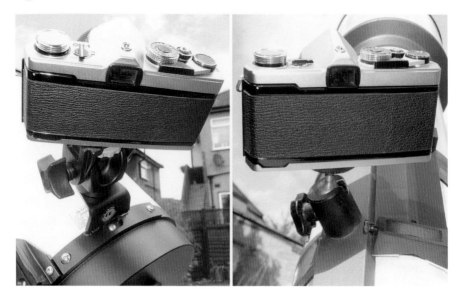

Figure 12.11. Olympus OM1 mounted on camera adaptors: SC-8 (left), AR-6 (right).

The LX90 telescope mount also has an AUX port. Meade's APM converts the port to allow the Autostar to control a motor focuser or illuminated reticle eyepiece, so it could be possible to connect it to the AUX port of an LXD55 or LXD75 mount. The motor focuser however, has to be Meade's own brand for the Autostar to operate it correctly (Meade #1206 Electric Focuser), and will only attach to the focuser of the LXD SC-8 model.

The APM as has a CCD port which is supposed to connect a compatible CCD to it, and provide some active tracking to the telescope mount. I have not known of anyone who has successfully auto-guided the telescope via a device connected to the CCD port of an APM on the LXD mount. It is best to use the Autostar serial interface port to successfully carry out auto-guiding via an LPI or similar device.

Piggyback Camera Mounts

This accessory allows a camera to be mounted on top of the OTA. Standard camera tripod type heads can be bought from most camera stockists, and retro fitted via a headless bolt to one of the slip rings. More dedicated types are available from most astronomical stockists.

LXD Carry Handle

The handle bolts directly to the OTA via the two threaded holes on the slip rings. Available from ScopeStuff (LXD55.com).

Figure 12.12. Autostar handset holder.

Autostar Placeholder

Simple Velcro straps to fit the Autostar handset to the telescope mount can be purchased from most hardware outlets. One side is placed on the back of the Autostar, whilst the other side placed on the base of the telescope. A piece of Velcro should be placed on each tripod leg, so that the Autostar can be accessed from all sides of the telescope.

Also available, is a metal handset holder for the LXD75 (Telescope House UK). The holder uses one of eyepiece holes on the LXD75 tripod mount (Figure 12.12). The handset holder is also used for other telescopes such as the Meade LX200 and the ETX.

Telescope Covers and Cases

Telescope covers help keep the telescope free from dust when it is not in use. They are available from most telescope stockists.

Telescope cases are ideal for transporting the telescope. Cases from JMI cater for all LXD OTAs in the series, including cases for the mounts, tripods and even the counterweights (Figure 12.13).

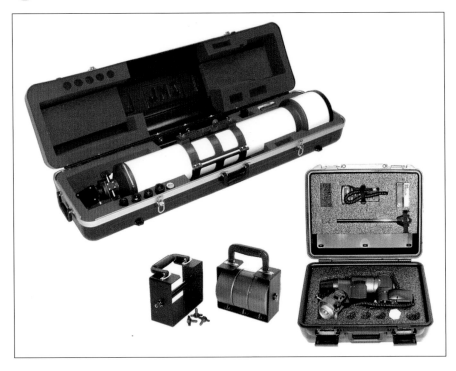

Figure 12.13. OTA telescope and tripod cases from JMI. Image courtesy of Jims Mobile.

Summary

As you can see, there are a whole host of accessories for the LXD series of telescopes. Budget may limit you to what you can purchase. Some accessories are luxury nice-to-haves, rather than improving the actual quality of observing. However, they all aid in the overall enjoyment of the hobby.

CHAPTER THIRTEEN

Where Did It All Go Wrong?

As with any mechanical or electronic device, telescopes can go wrong from time to time. When problems do occur, they tend to happen either, when first out of the box or at inconvenient times during an observing session. Perhaps a great observing opportunity is missed as a result of spending hours trying to fix the problem.

This chapter lists just some of the more common problems encountered by LXD owners. The list is not exhaustive by any means.

Internet websites, such as LXD55.com, LXD55 web pages of Mike Weasner's Mighty ETX Website and both Yahoo and MSN LXD user groups, provides volumes of information relating to problems with the LXD and Autostar in particular (see Appendix D). If you have a problem I would recommend searching the websites and forums first, before attempting to solve the problem by yourself.

However, if the telescope is still under warranty, then some actions suggested in this chapter may invalidate that warranty. Telescope warranties cover a variety of electrical and mechanical parts, so if you think the problem is serious and requires a mechanical repair, then you will need to contact your local stockist or Meade directly. See appendix D for Meade's contact address.

Alignment Stars are not in the FOV During Goto Setup Procedure – Alignment Failure is Displayed on the Autostar
(Chapter 5)

Probable Cause and Solution

Check the following Autostar parameters:

(A) Location
(B) Date and Time
(C) Polar alignment accuracy

The positions of the stars relative to the observer change from day to day (as discussed in Chapter 2). If the date and time is incorrectly set, then the Autostar software will think that the star is in a different position to where it actually is, so will slew to that incorrect position. If the misalignment stars appear to be the same distance offset from their actual positions, then it's likely that the date or time has been set incorrectly. Daylight saving time (DST) flag, if set incorrectly, can also cause this type of problem.

If problems are still occurring then make a note of the position of the alignment stars and follow the suggestions outlined in Chapter 5.

If the telescope slews accurately to the first star but the second star is not seen in the FOV, then the polar alignment needs to be checked (implies a probable cone error).

And finally if all else fails, perform one or more of the following:

–Calibrate and retrain the RA and Dec motors
–Perform a full Reset (A retrain of the motors is required after a Reset)
–Upgrade to the latest version of the Autostar firmware.
–Check the telescope is operating at full power (if using power cells).

Objects not Found in the FOV after a Goto is Performed
(Chapter 5)

Probable Cause and Solution

Refer in the first instance to the solution for the problem, discussed previously.

Every few hours perform the two or three star alignment process, to reduce cone error and maintain Goto accuracy.

The RA or Dec Axes Do not Turn When Autostar Arrows Keys are Pressed
(Chapter 4)

Probable Cause and Solution

The possible causes are primarily the following:

A. Motor gear is slipping against one of the axes wheels.
B. The cogs located on the side of the motors are slipping on their shafts.

See Chapter 4 for motor and cog adjustment details.

Autostar Displays 'Motor Unit Failure'
(Chapter 4)

Probable Cause and Solution

As a result of excessive binding between the motor and one of the axes, an unexpected signal was sent to the Autostar from one of the motor's digital encoders.

To correct the fault, slew the axis momentarily in 1 or 2 second bursts, using the Autostar arrow keys on a speed setting 5 or slower. If the motor noise indicates it is straining, and one or both the axes does not appear to be turning, stop the slewing immediately to prevent possible damage. In this case the motors will require re-aligning (see Chapter 4).

If the motors sound normal and the axis is turning as expected, run the 'calibrate motor' utility to attempt to clear the fault.

Constant motor failures could be a sign of something more serious, implying the motor may require servicing or replacing.

Poor Tracking and Backlash Problems
(Chapters 4 and 11)

Probable Cause and Solution

If the Motor units are not correctly aligned with the axes, there will be problems with backlash and tracking. Methods for reducing backlash are described in detail in Chapter 4.

Another method for ensuring smooth tracking is by slewing the axes at max speed setting. Several revolutions should be completed, to ensure the motor unit worm gear fully binds into every groove of the axes wheel circumference. This should smooth out any mechanical defects in the wheels and hence, improve the overall tracking.

A full overhaul and re-grease of the motor unit elements and axes wheels also vastly improves tracking, e.g. Hypertune/Supercharge (see Chapter 11).

Text on the Autostar Screen Appears Blurry and Slow to Display

Probable Cause and Solution

Check the power status of the battery. This can be done via the Autostar, if it is still partially useable.

The Battery Status can be displayed by holding down the Mode key for more than 2 seconds. Then, use the menu keys to scroll through the options until the status is shown (Appendix C).

A battery alarm function on the Autostar can also be set to inform you when the battery power reaches a particular level.

> Select Item → Utilities → Battery Alarm

Check the brightness and contrast settings of the Autostar display. Low brightness and contrast settings makes the text appear slow to display.

The brightness and contrast of the Autostar display can be adjusted through the Autostar menu options;

> Select Item → Utilities → Brightness Adj.

> Select Item → Utilities → Contrast Adj.

Autostar Unexpectedly Reboots

Probable Cause and Solution

Several causes contribute to this problem.

a. The power lead connecting the mains or battery power supply to the control panel is faulty, or the power plug fits loosely in the power socket of the control panel. In both cases, shaking the lead causes the Autostar to power itself off and on intermittently. Replace the power lead if it is faulty. However, if the power plug is the probable cause, before replacing the plug or lead, try widening the central power pin of the power socket with a small screwdriver to ensure a tighter fit of the power plug into the power socket.
b. The lead connecting the Autostar handset with the control panel is faulty. Other symptoms of a faulty lead includes, motors unexpectedly activating and turning the axes at full speed when the lead is shaken and, the Autostar unexpectedly performs a full Reset. The only solution is to replace the lead.
c. The HBX socket on the Autostar is faulty. There are two possible causes.

Where Did It All Go Wrong

Figure 13.1. Autostar socket PCB joints.

i. Two or more of the pins in the socket (or in the control panel HBX socket) are pushed together, causing a short circuit. Use a small screwdriver to gently separate the pins that are involved, (ensure power is switched off).

ii. Another possibility is that the internal soldered joints that hold the 8 pin socket in place on the internal circuit board are loose (known as a dry joint). This generally happens as a result of the socket not being well attached to the internal circuit board. When pressure is applied to the socket, such as the lead being excessively pulled upon, the soldered joints have to manage the force of the pull and break away from the circuit board joints.

If the telescope is still under warranty then the Autostar can be replaced or repaired by Meade. However, if the warranty has expired, then you can perform a little DIY on the handset yourself and repair the socket by re-soldering the joints (Figure 13.1).

If the Autostar lead is left constantly connected during storage it can put a force on the socket. Therefore it is advisable to disconnect the lead when the Autostar is not in use.

The Arrow Markers Located on the Side of Each Axis are not Aligned in Polar Home Position
(Chapter 4)

Probable Cause and Solution

The arrow markers assist in aligning the mount for Polar Home Position (Figure 13.2). The arrows are held on only with adhesive tape and are easily removed for repositioning. Some owners tend not to rely on them and use the setting circles instead.

A User's Guide to the Meade LXD55 and LXD75 Telescopes

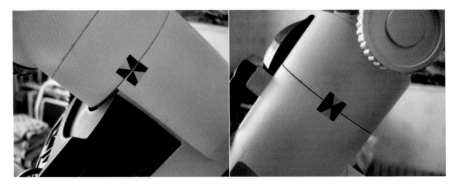

Figure 13.2. Arrow markers on RA (left) and Dec axes (right).

Summary

Some owners have been fortunate to have never had a problem with their LXD telescope. Others have not been so lucky and had to send their telescope back to Meade for repair. Even I was not immune from problems with both my LXD telescopes.

There is such a vast amount of information available to help you if you encounter a problem. Don't despair! There is always help at hand.

APPENDIX A

Lists and Charts of Autostar Named Stars

Table A.1 provides a list of named stars that are stored in the Autostar database. Following the list, there are constellation charts which show where the stars are located. The names are in alphabetical order along with their Latin designation (see Appendix B for complete list of constellations). Names in brackets ( ) in the table denote a different spelling to one that is known in the list.

The star's co-ordinates are set to the same as accuracy as the Autostar co-ordinates i.e. the RA or Dec 'sec' values are omitted.

Autostar option:

Select Item: Object → Star → Named

Appendix A

Table A.1. Autostar Named Star List

Named Star	Fig. Ref.	Latin Designation	RA Hr	RA Min	Dec Deg	Dec Min	Mag
Acamar	A5	Theta Eridanus	2	58.2	−40	18	3.2
Achernar	A5	Alpha Eridanus	1	37.6	−57	14	0.4
Acrux	A4	Alpha Crucis	12	26.5	−63	05	1.3
Adara	A2	Epsilon Canis Majoris	6	58.6	−28	58	1.5
Albireo	A4	Beta Cygni	19	30.6	++27	57	3.0
Alcor	A10	80 Ursae Majoris	13	25.2	+54	59	4.0
Alcyone	A9	Eta Tauri	3	47.4	+24	06	2.8
Aldebaran	A9	Alpha Tauri	4	35.8	+16	30	0.8
Alderamin	A3	Alpha Cephei	21	18.5	+62	35	2.4
Algenib	A7	Gamma Pegasi	0	13.2	+15	11	2.8
Algieba (Algeiba)	A6	Gamma Leonis	10	19.9	+19	50	2.6
Algol	A8	Beta Persei	3	8.1	+40	57	2.1
Alhena	A5	Gamma Geminorum	6	37.6	+16	23	1.9
Alioth	A10	Epsilon Ursae Majoris	12	54.0	+55	57	1.7
Alkaid	A10	Eta Ursae Majoris	13	47.5	+49	18	1.8
Almaak (Almach)	A1	Gamma Andromedae	2	3.8	+42	19	2.2
Alnair	A6	Alpha Gruis	22	8.2	−46	57	1.7
Alnath (Elnath)	A9	Beta Tauri	5	26.2	+28	36	1.6
Alnilam	A7	Epsilon Orionis	5	36.2	−01	12	1.7
Alnitak	A7	Zeta Orionis	5	40.7	−01	56	2.0
Alphard	A6	Alpha Hydrae	9	27.5	−08	39	1.9
Alphekka	A4	Alpha Coronae Borealis	15	34.6	+26	42	2.2
Alpheratz	A1	Alpha Andromedae	0	8.3	+29	05	2.0
Alshain	A1	Beta Aquilae	19	55.2	+06	24	3.7
Altair	A1	Alpha Aquilae	19	50.7	+08	52	0.7
Ankaa	A8	Alpha Phoenicis	0	26.2	−42	18	2.3
Antares	A9	Alpha Scorpii	16	29.4	−26	25	1.0
Arcturus	A2	Alpha Boötis	14	15.6	+19	10	0.1
Arneb	A7	Alpha Leporis	5	32.7	−17	49	2.5
Bellatrix	A7	Gamma Orionis	5	25.1	+06	20	1.6
Betelgeuse	A7	Alpha Orionis	5	55.1	+07	24	0.5
Canopus	A3	Alpha Carinae	6	23.9	−52	41	−0.6
Capella	A2	Alpha Aurigae	5	16.6	+45	59	0.0
Castor	A5	Alpha Geminorum	7	34.6	+31	53	1.5
Cor Caroli	A2	Alpha Canum Venaticorum	12	56.0	+38	19	2.8
Deneb	A4	Alpha Cygni	20	41.4	+45	16	1.3
Denebola	A6	Beta Leonis	11	49.0	+14	34	2.1
Diphda	A4	Beta Ceti	0	43.5	−17	59	2.0
Dubhe	A10	Alpha Ursae Majoris	11	3.7	+61	45	1.8
Enif	A7	Epsilon Pegasi	21	44.1	+09	52	2.3
Etamin	A5	Gamma Draconis	17	56.6	+51	29	2.2
Fomalhaut	A8	Alpha Piscis Austrini	22	57.6	−29	37	1.2
Hadar	A3	Beta Centauri	14	3.8	−60	22	0.6
Hamal	A1	Alpha Arietis	2	7.1	+23	27	2.0
Izar	A2	Epsilon Boötis	14	44.9	+27	04	2.5

(continued)

Appendix A

Table A.1. (Continued)

Named Star	Fig. Ref.	Latin Designation	RA Hr	RA Min	Dec Deg	Dec Min	Mag
Kaus Australis	A8	Epsilon Sagittarii	18	24.1	−34	23	1.8
Kocab (Kochab)	A10	Beta Ursae Minoris	14	50.7	+74	09	2.0
Markab	A7	Alpha Pegasi	23	4.7	+15	12	2.4
Megrez	A10	Delta Ursae Majoris	12	15.4	+57	01	3.3
Menkar	A4	Alpha Ceti	3	2.2	+04	05	2.5
Merak	A10	Beta Ursae Majoris	11	1.8	+56	22	2.3
Mintaka	A7	Delta Orionis	5	32.0	−00	17	2.2
Mira	A4	Omicron Ceti	2	19.3	−02	58	6.5
Mirach	A1	Beta Andromedae	1	9.7	+35	37	2.0
Mirphak	A8	Alpha Persei	3	24.3	+49	51	1.8
Mizar	A10	Zeta Ursae Majoris	13	23.9	+54	55	2.2
Nihal	A7	Beta Leporis	5	28.2	−20	45	2.8
Nunki	A8	Sigma Sagittarii	18	55.2	−26	17	2.0
Phad	A10	Gamma Ursae Majoris	11	53.8	+53	41	2.4
Polaris	A10	Alpha Ursae Minoris	2	31.8	+89	15	2.0
Pollux	A5	Beta Geminorum	7	45.3	+28	01	1.2
Procyon	A2	Alpha Canis Minoris	7	39.3	+05	13	0.4
Rasalgethi	A6	Alpha Herculis	17	14.6	+14	23	3.4
Rasalhague	A7	Alpha Ophiuchi	17	34.9	+12	33	2.0
Regulus	A6	Alpha Leonis	10	8.3	+11	58	1.4
Rigel	A6	Beta Orionis	5	14.5	−08	12	0.2
Sadalmelik	A1	Alpha Aquarii	22	5.7	−00	19	2.9
Saiph	A7	Kappa Orionis	5	47.7	−09	40	2.0
Scheat	A7	Beta Pegasi	23	3.7	+28	04	2.4
Shaula	A9	Lambda Scorpii	17	33.6	−37	06	1.6
Shedir (Shedar)	A3	Alpha Cassiopeiae	0	40.5	+56	32	2.2
Sirius	A2	Alpha Canis Majoris	6	45.1	−16	42	−1.4
Spica	A10	Alpha Viriginis	13	25.1	−11	09	1.0
Tarazed	A1	Gamma Aquilae	19	46.2	+10	36	2.7
Thuban	A5	Alpha Draconis	14	4.3	+64	22	3.6
Unukalhai	A9	Alpha Serpentis	15	44.2	+06	25	2.6
Vega	A7	Alpha Lyrae	18	36.9	+38	47	0.0
Vindemiatrix	A10	Epsilon Virginis	13	2.1	+10	57	2.8

(Star charts to follow)

Appendix A

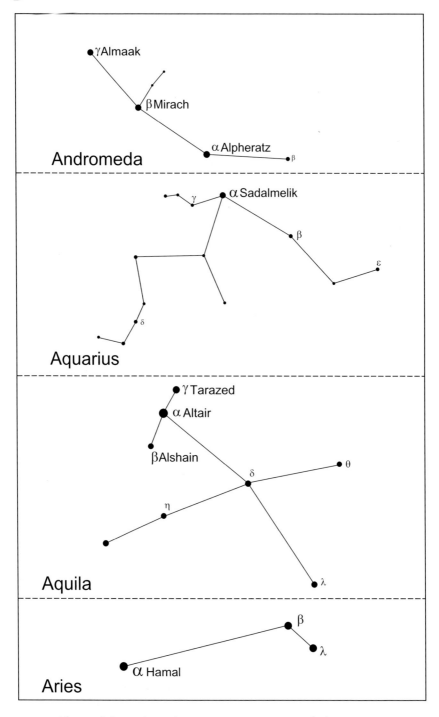

Figure A.1. Andromeda to Aries. Image courtesy of Alan Marriott.

Appendix A

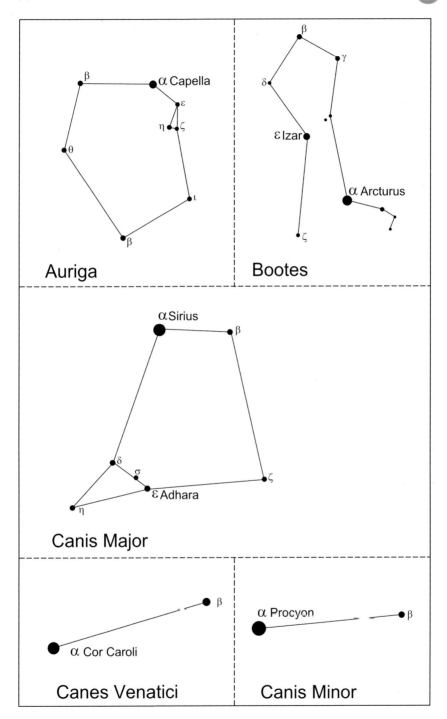

Figure A.2. Auriga to Canis Minor. Image courtesy of Alan Marriott.

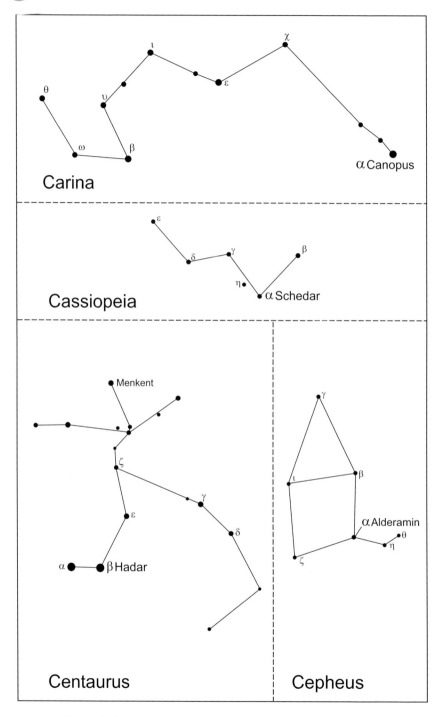

Figure A.3. Carina to Cepheus. Image courtesy of Alan Marriott.

Appendix A

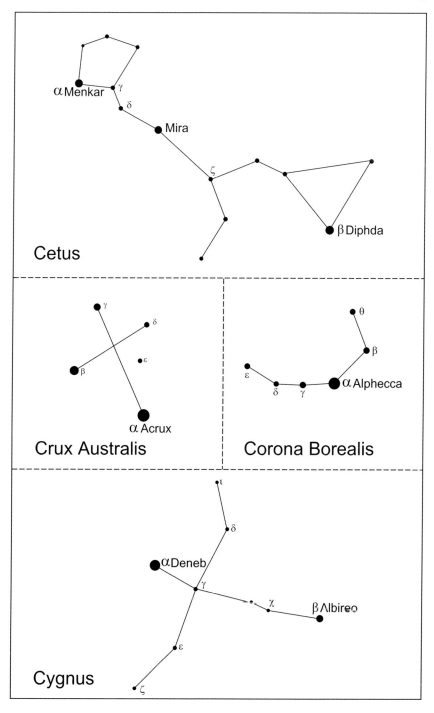

Figure A.4. Cetus to Cygnus. Image courtesy of Alan Marriott.

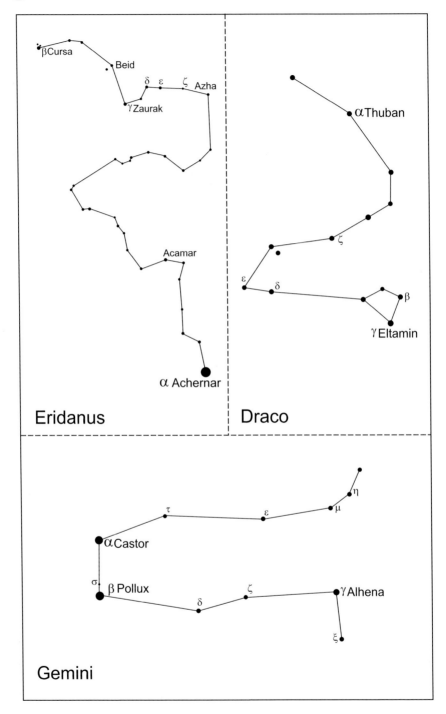

Figure A.5. Eridanus to Gemini. Image courtesy of Alan Marriott.

Appendix A

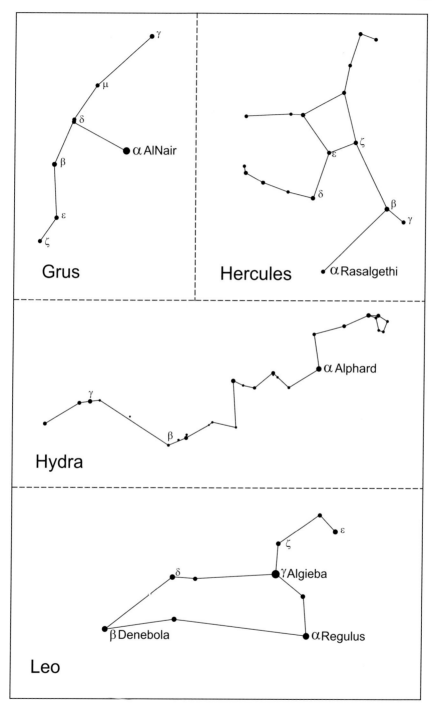

Figure A.6. Grus to Leo. Image courtesy of Alan Marriott.

Appendix A

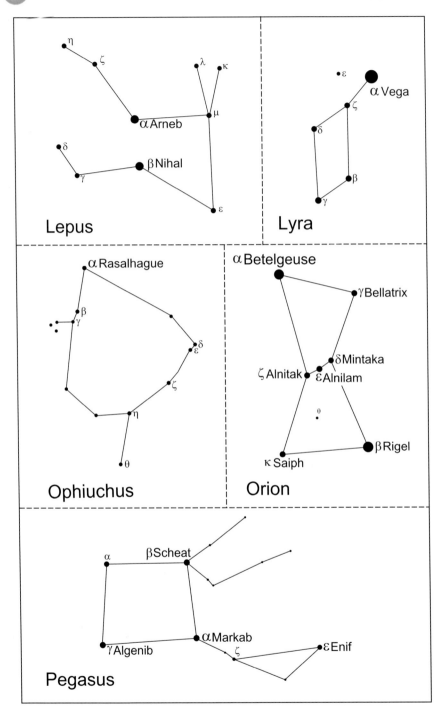

Figure A.7. Lepus to Pegasus. Image courtesy of Alan Marriott.

Appendix A

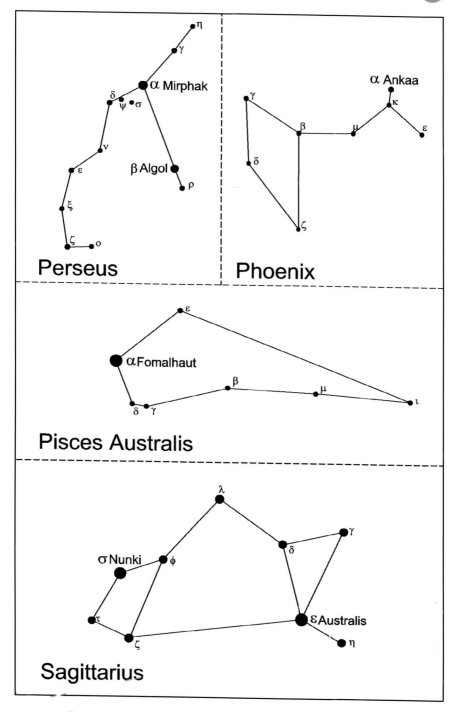

Figure A.8. Perseus to Sagittarius. Image courtesy of Alan Marriott.

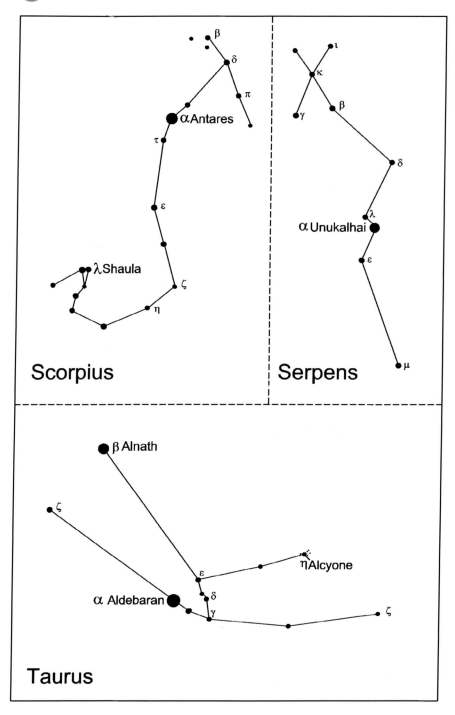

Figure A.9. Scorpius to Taurus. Image courtesy of Alan Marriott.

Appendix A

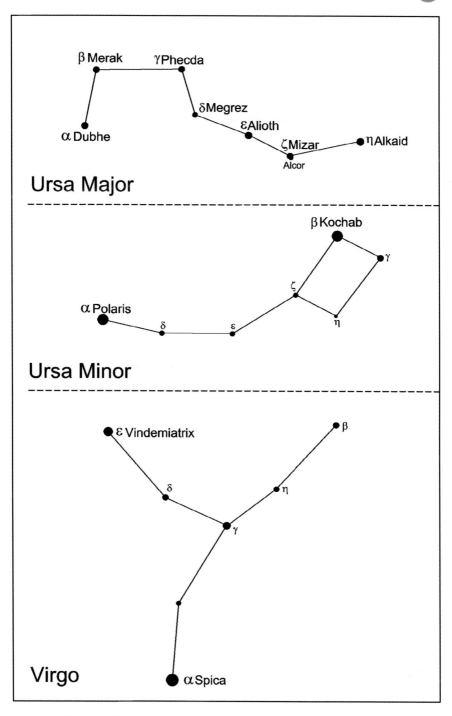

Figure A.10. Ursa Major to Virgo. Image courtesy of Alan Marriott.

APPENDIX B

Object Lists

B1. Autostar Constellation List

Autostar option:

Select Item: Object → Constellation

Table B.1. Autostar Constellation List.

Constellation	Abbreviation	Latin Genitive
Andromeda	And	Andromedae
Antlia	Ant	Antliae
Apus	Aps	Apodis
Aquarius	Aqr	Aquarii
Aquila	Aql	Aquilae
Ara	Ara	Arae
Aries	Ari	Arietis
Auriga	Aur	Aurigae
Boötes	Boo	Boötis
Caelum	Cae	Caeli
Camelopardalis	Cam	Camelopardalis
Cancer	Cnc	Cancri
Canes Venatici	CVn	Canum Venaticorum
Canis Major	CMa	Canis Majoris
Canis Minor	CMi	Canis Minoris

(*continued*)

Table B.1. (*Continued*)

Constellation	Abbreviation	Latin Genitive
Capricornus	Cap	Capricorni
Carina	Car	Carinae
Cassiopeia	Cas	Cassiopeiae
Centaurus	Cen	Centauri
Cepheus	Cep	Cephei
Cetus	Cet	Ceti
Chamaeleon	Cha	Chamaeleontis
Circinus	Cir	Circini
Columba	Col	Columbae
Coma Berenices	Com	Comae Berenices
Corona Australis	CrA	Coronae Australis
Corona Borealis	CrB	Coronae Borealis
Corvus	Crv	Corvi
Crater	Crt	Crateris
Crux	Cru	Crucis
Cygnus	Cyg	Cygni
Delphinus	Del	Delphini
Dorado	Dor	Doradus
Draco	Dra	Draconis
Equuleus	Equ	Equulei
Eridanus	Eri	Eridani
Fornax	For	Fornacis
Gemini	Gem	Geminorum
Grus	Gru	Gruis
Hercules	Her	Herculis
Horologium	Hor	Horologii
Hydra	Hya	Hydrae
Hydrus	Hyi	Hydri
Indus	Ind	Indi
Lacerta	Lac	Lacertae
Leo	Leo	Leonis
Leo Minor	LMi	Leonis Minoris
Lepus	Lep	Leporis
Libra	Lib	Librae
Lupus	Lup	Lupi
Lynx	Lyn	Lyncis
Lyra	Lyr	Lyrae
Mensa	Men	Mensae
Microscopium	Mic	Microscopii
Monoceros	Mon	Monocerotis
Musca	Mus	Muscae
Norma	Nor	Normae
Octans	Oct	Octantis
Ophiuchus	Oph	Ophiuchi
Orion	Ori	Orionis

(*continued*)

Appendix B

Table B.1. (*Continued*)

Constellation	Abbreviation	Latin Genitive
Pavo	Pav	Pavonis
Pegasus	Peg	Pegasi
Perseus	Per	Persei
Phoenix	Phe	Phoenicis
Pictor	Pic	Pictoris
Pisces	Psc	Piscium
Piscis Austrinus	PsA	Piscis Austrini
Puppis	Pup	Puppis
Pyxis	Pyx	Pyxidis
Reticulum	Ret	Reticuli
Sagitta	Sge	Sagittae
Sagittarius	Sgr	Sagittarii
Scorpius	Sco	Scorpii
Sculptor	Scl	Sculptoris
Scutum	Sct	Scuti
Serpens	Ser	Serpentis
Sextans	Sex	Sextantis
Taurus	Tau	Tauri
Telescopium	Tel	Telescopii
Triangulum	Tri	Trianguli
Triangulum Australe	TrA	Trianguli Australis
Tucana	Tuc	Tucanae
Ursa Major	UMa	Ursae Majoris
Ursa Major	UMi	Ursae Minoris
Vela	Vel	Velorum
Virgo	Vir	Virginis
Volans	Vol	Volantis
Vulpecula	Vul	Vulpeculae

B2. Autostar Messier Objects List

Autostar option:

> Select Item: Object → Deep Sky → Messier Objects

Press the '?' key on the handset for more details about the object.

Table B.2. Autostar Messier Objects List.

Messier	NGC/IC ID	Common Name	Type	Const.	RA Hr	RA Min	DEC Deg	DEC Min	Mag
1	NGC 1952	Crab Nebula	SR	Tau	5	34.5	+22	1	8.4
2	NGC 7089		GC	Aqr	21	33.5	−00	49	6.5
3	NGC 5272		GC	CVn	13	42.2	+28	23	6.2
4	NGC 6121		GC	Sco	16	23.6	−26	32	5.6
5	NGC 5904		GC	Ser	15	18.6	+02	5	5.6
6	NGC 6405	Butterfly Cluster	OC	Sco	17	40.1	−32	13	5.3
7	NGC 6475	Ptolemy's Cluster	OC	Sco	17	53.9	−34	49	4.1
8	NGC 6523	Lagoon Nebula	DN	Sgr	18	3.8	−24	23	6
9	NGC 6333		GC	Oph	17	19.2	−18	31	7.7
10	NGC 6254		GC	Oph	16	57.1	−04	6	6.6
11	NGC 6705	Wild Duck Cluster	OC	Sct	18	51.1	−06	16	6.3
12	NGC 6218		GC	Oph	16	47.2	−01	57	6.7
13	NGC 6205	Hercules Cluster	GC	Her	16	41.7	+36	28	5.8
14	NGC 6402		GC	Oph	17	37.6	−03	15	7.6
15	NGC 7078	Pegasus Cluster	GC	Peg	21	30	+12	10	6.2
16	NGC 6611	Eagle Nebula Cluster	OC	Ser	18	18.8	−13	47	6.4
17	NGC 6618	Omega, Swan	DN	Sgr	18	20.8	−16	11	7
18	NGC 6613		OC	Sgr	18	19.9	−17	8	7.5
19	NGC 6273		GC	Oph	17	2.6	−26	16	6.8
20	NGC 6514	Trifid Nebula	DN	Sgr	18	2.3	−23	2	9
21	NGC 6531		OC	Sgr	18	4.6	−22	30	6.5
22	NGC 6656		GC	Sgr	18	36.4	−23	54	5.1
23	NGC 6494		OC	Sgr	17	56.8	−19	1	6.9
24	NGC 6603	Milky Way Patch	SC	Sgr	18	18.4	−18	25	4.6
25	IC 4725		OC	Sgr	18	31.7	−19	14	6.5
26	NGC 6694		OC	Sct	18	45.2	−09	24	8
27	NGC 6853	Dumbbell Nebula	PN	Vul	19	59.6	+22	43	7.4
28	NGC 6626		GC	Sgr	18	24.5	−24	52	6.8
29	NGC 6913		OC	Cyg	20	23.9	+38	32	7.1
30	NGC 7099		GC	Cap	21	40.4	−23	11	7.2
31	NGC 224	Andromeda Galaxy	SG	And	0	42.7	+41	16	3.4
32	NGC 221	Satellite of M31	EG	And	0	42.7	+40	52	8.1
33	NGC 598	Pinwheel Galaxy	SG	Tri	1	33.9	+30	39	5.7
34	NGC 1039		OC	Per	2	42	+42	47	5.5
35	NGC 2168		OC	Gem	6	8.9	+24	20	5.3
36	NGC 1960		OC	Aur	5	36.1	+34	8	6.3
37	NGC 2099		OC	Aur	5	52.4	+32	33	6.2
38	NGC 1912		OC	Aur	5	28.7	+35	50	7.4
39	NGC 7092		OC	Cyg	21	32.2	+48	26	5.2
40	Winecke 4		DS	UMa	12	22.2	+58	5	9/9.3
41	NGC 2287		OC	CMi	6	47	−20	44	4.6
42	NGC 1976	Great Orion Nebula	DN	Ori	5	35.4	−05	27	4
43	NGC 1982	de Mairan's Nebula	DN	Ori	5	35.6	−05	16	9
44	NGC 2632	Beehive Cluster, Praesepe	OC	Cnc	8	40.1	+19	59	3.7
45	None	Pleiades, Seven Sisters	OC	Tau	3	47	+24	7.2	1.6
46	NGC 2437		OC	Pup	7	41.8	−14	49	6
47	NGC 2422		OC	Pup	7	36.6	−14	30	5.2
48	NGC 2548		OC	Hya	8	13.8	−05	48	5.5

(continued)

Appendix B

Table B.2. (*Continued*)

Messier	NGC/IC ID	Common Name	Type	Const.	RA Hr	RA Min	DEC Deg	DEC Min	Mag
49	NGC 4472		EG	Vir	12	29.8	+08	0	8.4
50	NGC 2323		OC	Mon	7	3.2	−08	20	6.3
51	NGC 5194	Whirlpool Galaxy	SG	CVn	13	29.9	+47	12	8.4
52	NGC 7654		OC	Cas	23	24.2	+61	35	7.3
53	NGC 5024		GC	Com	13	12.9	+18	10	7.6
54	NGC 6715		GC	Sgr	18	55.1	−30	29	7.6
55	NGC 6809		GC	Sgr	19	40	−30	58	6.3
56	NGC 6779		GC	Lyr	19	16.6	+30	11	8.3
57	NGC 6720	Ring Nebula	PN	Lyr	18	53.6	+33	2	8.8
58	NGC 4579		SG	Vir	12	37.7	+11	49	9.7
59	NGC 4621		EG	Vir	12	42	+11	39	9.6
60	NGC 4649		EG	Vir	12	43.7	+11	33	8.8
61	NGC 4303		SG	Vir	12	21.9	+04	28	9.7
62	NGC 6266		GC	Oph	17	1.2	−30	7	6.5
63	NGC 5055	Sunflower Galaxy	SG	CVn	13	15.8	+42	2	8.6
64	NGC 4826	Blackeye Galaxy	SG	Com	12	56.7	+21	41	8.5
65	NGC 3623		SG	Leo	11	18.9	+13	5	9.3
66	NGC 3627		SG	Leo	11	20.2	+12	59	8.9
67	NGC 2682		OC	Cnc	8	50.4	+11	49	6.1
68	NGC 4590		GC	Hya	12	39.5	−26	45	7.8
69	NGC 6637		GC	Sgr	18	31.4	−32	21	7.6
70	NGC 6681		GC	Sgr	18	43.2	−32	18	7.9
71	NGC 6838		GC	Sge	19	53.8	+18	47	8.3
72	NGC 6981		GC	Aqr	20	53.5	−12	32	9.3
73	NGC 6994		Ast	Aqr	20	59	−12	38	9
74	NGC 628		SG	Psc	1	36.7	+15	47	9.4
75	NGC 6864		GC	Sgr	20	6.1	−21	55	8.5
76	NGC 650	Little Dumbbell	PN	Per	1	42.3	+51	34	10.1
77	NGC 1068	Cetus A	SG	Cet	2	42.7	+00	1	8.9
78	NGC 2068		DN	Ori	5	46.7	+00	3	8.3
79	NGC 1904		GC	Lep	5	24.5	−24	33	7.7
80	NGC 6093		GC	Sco	16	17	−22	59	7.3
81	NGC 3031	Bode's Galaxy	SG	UMa	9	55.6	+69	4	6.9
82	NGC 3034	Cigar Galaxy	IG	UMa	9	55.8	+69	41	8.4
83	NGC 5236	Southern Pinwheel	SG	Hya	13	37	−29	52	7.6
84	NGC 4374		LG	Vir	12	25.1	+12	53	9.1
85	NGC 4382		LG	Com	12	25.4	+18	11	9.1
86	NGC 4406		LG	Vir	12	26.2	+12	57	8.9
87	NGC 4486	Virgo A	EG	Vir	12	30.8	+12	24	8.6
88	NGC 4501		SG	Com	12	32	+14	25	9.6
89	NGC 4552		EG	Vir	12	35.7	+12	33	9.8
90	NGC 4569		SG	Vir	12	36.8	+13	10	9.5
91	NGC 4548		SG	Com	12	35.4	+14	30	10.2
92	NGC 6341		GC	Her	17	17.1	+43	8	6.4
93	NGC 2447		OC	Pup	7	44.6	−23	52	6
94	NGC 4736		SG	CVn	12	50.9	+41	7	8.2
95	NGC 3351		SG	Leo	10	44	+11	42	9.7
96	NGC 3368		SG	Leo	10	46.8	+11	49	9.2

(*continued*)

Appendix B

Table B.2. (*Continued*)

Messier	NGC/IC ID	Common Name	Type	Const.	RA Hr	RA Min	DEC Deg	DEC Min	Mag
97	NGC 3587	Owl Nebula	PN	UMa	11	14.8	+55	1	9.9
98	NGC 4192		SG	Com	12	13.8	+14	54	10.1
99	NGC 4254		SG	Com	12	18.8	+14	25	9.9
100	NGC 4321		SG	Com	12	22.9	+15	49	9.3
101	NGC 5457	Pinwheel Galaxy	SG	UMa	14	3.2	+54	21	7.9
102	NGC 5866	Spindle Galaxy	LG	Dra	15	6.5	+55	46	9.9
103	NGC 581		OC	Cas	1	33.2	+60	42	7.4
104	NGC 4594	Sombrero Galaxy	SG	Vir	12	40	−11	37	8
105	NGC 3379		EG	Leo	10	47.8	+12	35	9.3
106	NGC 4258		SG	CVn	12	19	+47	18	8.4
107	NGC 6171		GC	Oph	16	32.5	−13	3	7.9
108	NGC 3556		SG	UMa	11	11.5	+55	40	10
109	NGC 3992		SG	UMa	11	57.6	+53	23	9.8
110	NGC 205	Satellite of M31	EG	And	0	40.4	+41	41	8.5

Note: Ast – Star group/Asterism; DN – Diffuse Nebula; DS – Double Star; EG – Elliptical type Galaxy; GC – Globular Cluster; IG – Irregular Galaxy; LG – Lenticular type Galaxy; OC – Open Cluster; PN – Planetary Nebula; SC – Star Cloud; SG – Spiral type Galaxy; SR – Supernova Remnant.

B3. Autostar Caldwell Objects List

Autostar option:

Select Item: Object → Deep Sky → Caldwell Objects

Press the '?' key on the handset for more details about the object.

Table B.3. Autostar Caldwell Objects List.

Caldwell	NGC/IC	Type*	Const.	RA Hr	RA Min	DEC Deg	DEC Min	Mag	Description
1	188	OC	Cep	0	44.4	85	20	8.1	Cluster close to NCP
2	40	PN	Cep	0	13	72	32	11.6	
3	4236	SG	Dra	12	16.7	69	28	9.7	
4	7023	BN	Cep	21	1.8	68	12	6.8	Reflection Nebula
5	IC342	SG	Cam	3	46.8	68	6	9.2	
6	6543	PN	Dra	17	58.6	66	38	8.8	Cat's Eye Nebula
7	2403	SG	Cam	7	36.9	65	36	8.9	
8	559	OC	Cas	1	29.5	63	18	9.5	

(*continued*)

Appendix B

Table B.3. (*Continued*)

Caldwell	NGC/IC	Type*	Const.	RA Hr	RA Min	DEC Deg	DEC Min	Mag	Description
9	Sh2155	BN	Cep	22	56.8	62	37	7.7	Cave Nebula
10	663	OC	Cas	1	46	61	15	7.1	
11	7635	BN	Cas	23	20.7	61	12	7	Bubble Nebula
12	6946	SG	Cep	20	34.8	60	9	9.7	
13	457	OC	Cas	1	19.1	58	20	6.4	Owl/Phi Cas Cluster
14	869/884	DoubC	Per	2	20	57	8	4.3	Sword Handle
15	6826	PN	Cyg	19	44.8	50	31	9.8	Blinking Nebula
16	7243	OC	Lac	22	15.3	49	53	6.4	
17	147	EG	Cas	0	33.2	48	30	9.3	Remote Satellite of M31
18	185	EG	Cas	0	39	48	20	9.2	Remote Satellite of M31
19	IC5146	BN	Cyg	21	53.5	47	16	10	Cocoon Nebula
20	7000	BN	Cyg	20	58.8	44	20	6	North American Nebula
21	4449	IG	CVn	12	28.2	44	6	9.4	
22	7662	PN	And	23	25.9	42	33	9.2	
23	891	SG	And	2	22.6	42	21	9.9	
24	1275	SeyF.G	Per	3	19.8	41	31	11.6	Perseus A Radio Source
25	2419	GC	Lyn	7	38.1	38	53	10.4	Distant Globular Cluster
26	4244	SG	CVn	12	17.5	37	49	10.6	Whale Galaxy
27	6888	BN	Cyg	20	12	38	21	7.5	Crescent Nebula
28	752	OC	And	1	57.8	37	41	5.7	
29	5005	SG	CVn	13	10.9	37	3	9.8	
30	7331	SG	Peg	22	37.1	34	25	9.5	
31	IC405	BN	Aur	5	16.2	34	16	6	Flaming Star Nebula
32	4631	SG	CVn	12	42.1	32	32	9.3	
33	6992	SN	Cyg	20	56.4	31	43	–	East Veil Nebula
34	6960	SN	Cyg	20	45.7	30	43	–	West Veil Nebula
35	4889	EG	Com	13	0.1	27	59	11.4	
36	4559	SG	Com	12	36	27	58	9.8	
37	6885	OC	Vul	20	12	26	29	5.7	
38	4565	SG	Com	12	36.3	25	59	9.6	Needle Galaxy
39	2392	PN	Gem	7	29.2	20	55	9.9	Eskimo Nebula
40	3626	SG	Leo	11	20.1	18	21	10.9	
41	–	OC	Tau	4	27	16	0	1	Hyades
42	7006	GC	Del	21	1.5	16	11	10.6	
43	7814	SG	Peg	0	3.3	16	9	10.5	
44	7479	SG	Peg	23	4.9	12	19	11	
45	5248	SG	Boo	13	37.5	8	53	10.2	
46	2261	BN	Mon	6	39.2	8	44	10	Hubble's Variable Nebula
47	6934	GC	Del	20	34.2	7	24	0.9	
48	2775	SG	Can	9	10.3	7	2	10.3	
49	2237	BN	Mon	6	32.3	5	3	–	Rosette Nebula
50	2244	OC	Mon	6	32.4	4	52	4.8	
51	IC1613	IG	Cet	1	4.8	2	7	9	
52	4697	EG	Vir	12	48.6	–5	48	9.3	
53	3115	EG	Sex	10	5.2	–7	43	9.1	Spindle Galaxy
54	2506	OC	Mon	8	0.2	–10	47	7.6	

(*continued*)

Table B.3. (*Continued*)

Caldwell	NGC/IC	Type*	Const.	RA Hr	RA Min	DEC Deg	DEC Min	Mag	Description
55	7009	PN	Aqr	21	4.2	−11	22	8.3	Saturn Nebula
56	246	PN	Cet	0	47	−11	53	8	
57	6822	IG	Sgr	19	44.9	−14	48	9.3	Barnard's Galaxy
58	2360	OC	CMa	7	17.8	−15	37	7.2	
59	3242	PN	Hya	10	24.8	−18	38	8.6	Ghost of Jupiter
60	4038	SG	Crv	12	1.9	−18	52	11.3	The Antennae
61	4039	SG	Crv	12	1.9	−18	53	13	The Antennae
62	247	SG	Cet	0	47.1	−20	46	8.9	
63	7293	PN	Aqr	22	29.6	−20	48	6.5	Helix Nebula
64	2362	OC	CMa	7	18.8	−24	57	4.1	Tau Canis Major Cluster
65	253	SG	Scl	0	47.6	−25	17	7.1	Sculptor Galaxy
66	5694	GC	Hya	14	39.6	−26	32	10.2	
67	1097	SG	For	2	46.3	−30	17	9.2	
68	6729	BN	CrA	19	1.9	−36	57	9.7	R Corona Borealis Nebula
69	6302	PN	Sco	17	13.7	−37	6	12.8	Bug Nebula
70	300	SG	Scl	0	54.9	−37	41	8.1	
71	2477	OC	Pup	7	52.3	−38	33	5.8	
72	55	SG	Scl	0	14.9	−39	11	8.2	
73	1851	GC	Col	5	14.1	−40	3	7.3	
74	3132	PN	Vel	10	7.7	−40	26	8.2	
75	6124	OC	Sco	16	25.6	−40	40	5.8	
76	6231	OC	Sco	16	54	−41	48	2.6	
77	5128	Pec.G	Cen	13	25.5	−43	1	7	Centaurus A Radio Source
78	6541	GC	CrA	18	8	−43	42	6.6	
79	3201	GC	Vel	10	17.6	−46	25	6.7	
80	5139	GC	Cen	13	26.8	−47	29	3.6	Omega Centauri
81	6352	GC	Ara	17	25.5	−48	25	8.1	
82	6193	OC	Ara	16	41.3	−48	46	5.2	
83	4945	SG	Cen	13	5.4	−49	28	9.5	
84	5286	GC	Cen	13	46.4	−51	22	7.6	
85	IC2391	OC	Vel	8	40.2	−53	4	2.5	Omicron Vela Cluster
86	6397	GC	Ara	17	40.7	−53	40	5.6	
87	1261	GC	Hor	3	12.3	−55	13	8.4	
88	5823	OC	Cir	15	5.7	−55	36	7.9	
89	6087	OC	Nor	16	18.9	−57	54	5.4	S Normae Cluster
90	2867	PN	Car	9	21.4	−58	19	9.7	
91	3532	OC	Car	11	6.4	−58	40	3	
92	3372	BN	Car	10	43.8	−59	52	6.2	Eta Carinae Nebula
93	6752	GC	Pav	19	10.9	−59	59	5.4	
94	4755	OC	Cru	12	53.6	−60	20	4.2	Jewel Box Cluster
95	6025	OC	TrA	16	3.7	−60	30	5.1	
96	2516	OC	Car	7	58.3	−60	52	3.8	
97	3766	OC	Cen	11	36.1	−61	37	5.3	
98	4609	OC	Cru	12	42.3	−62	58	6.9	
99	–	DrkN	Cru	12	53	−63	0	–	Coalsack Dark Nebula
100	IC2994	OC	Cen	11	36.6	−63	2	4.5	Lambda Centaurus Cluster
101	6744	SG	Pav	19	9.8	−63	51	9	

(*continued*)

Appendix B

Table B.3. (*Continued*)

Caldwell	NGC/IC	Type*	Const.	RA Hr	RA Min	DEC Deg	DEC Min	Mag	Description
102	IC2602	OC	Car	10	43.2	−64	24	1.9	Theta Carina Cluster
103	2070	BN	Dor	5	38.7	−69	6	1	Tarantula Nebula in (LMC)
104	362	GC	Tuc	1	3.2	−70	51	6.6	
105	4833	GC	Mus	12	59.6	−70	53	7.3	
106	104	GC	Tuc	0	24.1	−72	5	4	47 Tucanae
107	6101	GC	Aps	16	25.8	−72	12	9.3	
108	4372	GC	Mus	12	25.8	−72	40	7.8	
109	3195	PN	Cha	10	9.5	−80	52	–	

Note: BN – Bright Nebula; DoubC – Double Cluster; GC – Globular Cluster; OC – Open Cluster; PN – Planetary Nebula; SG – Spiral type Galaxy; EG – Elliptical type Galaxy; IG – Irregular Galaxy; Pec.G – Peculiar Galaxy; SeyF.G – Seyfert Galaxy; SR – Supernova Remnant; DrkN – Dark Nebula.

B4. Annual Meteor Showers

Table B.4. Meteor Showers.

Meteor Shower	Location	Peak	Duration	Peak Rate	Description	Comet
Quadrantids	Bootes	Jan 2	1 day	85/hr	Many, Fast, Faint	C/1490 Y1*
Lyrids	Lyra	Apr 21	2 days	10/hr	Bright, Blue	Thatcher
Eta Aquarids	Aquarius	May 4	3 days	35/hr	Fast, Persistent trails	Halley
Delta Aquarids	Aquarius	May 27	7 days	20/hr	Slow, Long paths	Unknown
Perseids	Perseus	Aug 11	5 days	75/hr	Fast, fragmenting	Swift Tuttle
Orionids	Orion	Oct 20	2 days	25/hr	Fast, Persistent trails	Halley
Taurids	Taurus	Nov 2	20 days	20/hr	Slow, Brilliant	Encke
Leonids	Leo	Nov 16	5 days	25/hr	Fast, Persistent trails	Temple Tuttle
Geminids	Gemini	Dec 13	3 days	75/hr	White, bright, many	3200 Phaethon
Ursids	Ursa Min.	Dec 21	2 days	15/hr	White, some persistent	Tuttle**

*Minor Planet 2003EH_1 – to be confirmed.
**No direct correlation confirmed.

APPENDIX C

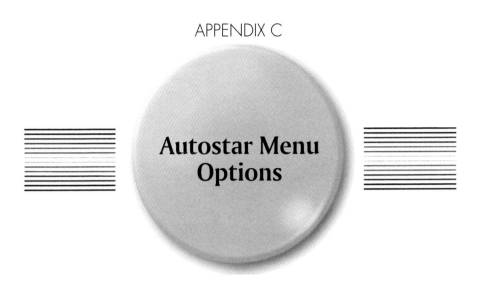

Autostar Menu Options

For Constellation menu list – See Appendix B.1.
For Glossary menu – see Autostar handset for explanation of terms A through Z.
To navigate through the menus, move across then down.

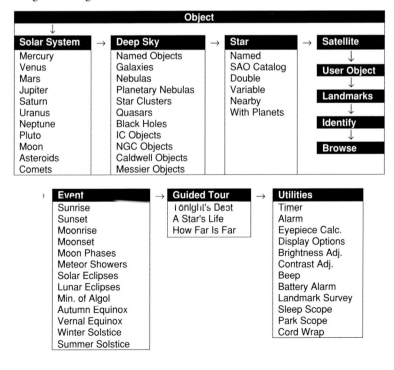

Appendix C

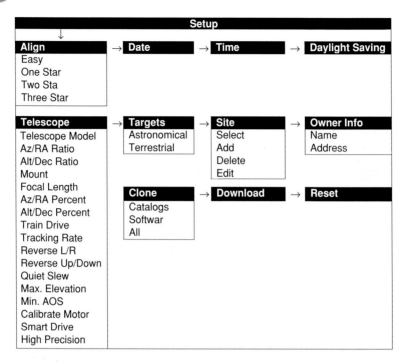

Autostar Status Display
(Hold the Mode Key down for more than two seconds):

Mode
RA/Dec
Alt/Az
Site/Date
Latitude
Longitude
Time/LST
Timer/Alarm
Battery Level

APPENDIX D

References and Further Reading

D.1 List of Useful Websites

LXD Telescope Websites

If you type LXD, LXD55 or LXD75 into any search engine you will see that there are a vast amount of websites dedicated to the telescope models. Some of the more popular ones are listed below.

MEADE USA

http://www.meade.com
Address:
Meade Instruments Corporation
6001 Oak Canyon
Irvine, CA 92618
(800) 626-3233 (U.S.A. Only)
+1(949) 451-1450
Fax: (949) 451-1460

    The Meade website and mail address. The website provides commercial information about the LXD telescope series. The site is where you can also download the latest Autostar handset software and Autostar update facility (ASU).
    http://www.meade.com/autostar/as_suite.html

LXD55.COM

http://LXD55.com. Conceived and managed by Richard Harris, the website started in 2001 as a Meade fan site. It has since grown to over 5000 visitors on average a month. The resources there include a vast knowledge-base, Astrophoto section (by members), forums, and much more. Being a member is free and there are currently 1500 worldwide members. The site also offers several LXD55 and LXD75 upgrade products there including the well-known HyperTune Service and Do-It-YourSelf HyperTune DVD and service kit. The HyperTune service was created in 2002, and has since become the standard in GEM upgrades which span the LXD55, LXD75, CG5, and the like.

LXD75.COM

http://LXD75.com. This was started by Richard Harris in 2004 as a daughter site of LXD55.com. It is a free site dedicated to Meade LXD75 owners and would-be owners. It has many of the same resources as LXD55.com and is growing in popularity every week. There, you can find owners sharing photos, discussing telescope problems, and becoming better acquainted with the telescope in general. To date LXD75.com has 400 worldwide members.

Richard also runs ScopeTrader.com (http://ScopeTrader.com) which is a free astronomy resource site for amateur astronomers to share photos, read and upload astronomy articles. There are astronomical gear reviews, and has a vast astronomy gear classifieds section.

Mike Weasner's LXD55/75 Site

http://www.weasner.com/lxd. To follow on from Mike's very successful Weasner's Mighty ETX website, a sister LXD55 site was created in 2002 when Mike purchased an LXD55 SC-8 telescope. In May 2006 he recently upgraded to an LXD75 SC-8.

The free site contains vast amounts of information on practically every aspect of the LXD series including Mike's personal experiences of using his telescopes, astrophotography galleries, mechanical upgrades, Autostar problems and general hints and tips. The site also contains a complete archive of every version of the Autostar software and patches to date. The ETX website contains information that is also applicable for the LXD telescope series. If you have a question about your LXD telescope then send an Email to Mike.

LXD Telescope Series Yahoo Telescope Groups

http://groups.yahoo.com/group/LXD55telescopes/
http://groups.yahoo.com/group/LXD75telescopes/

The LXD55 group was founded in September 2001 as a prelude to the LXD telescope before it was available to buy. This is an open forum for discussion of all Meade LXD55 Telescopes. It has over 3500 members. The LXD75 group was founded in May 2004 and has over 1000 members.

http://www.frappr.com/lxd55andlxd75owners

The frappr website provides a map of LXD55 and LXD75 users around the globe. Sponsored by LXD55.com.

Appendix D

Dr P. Clay Sherrod – Arkansas Sky Observatory

http://www.arksky.org/. This popular astronomy website has many useful hints and tips. It links to Dr Sherrod's Supercharge Service, which caters for a variety of telescopes including the LXD55/75 telescope series.

Other Websites and Resources

General Astronomy and Astronomy Home Pages:

Astronomy Thesaurus	http://msowww.anu.edu.au/library/thesaurus/
Astrophotos from George Tarsoudis	http://www.flickr.com/photos/80161946@N00/?saved=1
Cloudy Nights Telescope Reviews	http://www.cloudynights.com/
Meade 4M Community	http://www.meade4m.com/
Phuturespace. Home page of Martin Peston (UK)	http://www.phuturespace.com
Sunflower Astronomy. Home Page of David Kolb	http://www.sunflower-astronomy.com/
The Lunar Navigator Interactive Maps Of The Moon	http://www.lunarrepublic.com/atlas/index.shtml
Tom How's guide to Autoguiding with an LXD55 (UK)	http://astro.neutral.org/eq/autoguide.html
Wikipedia	http://www.wikipedia.com
Yahoo Astronomy Information	http://dir.yahoo.com/Science/astronomy

Meade LXD Distributors and Stockists:

Telescope House UK. Stockist and distributors of Meade products for the UK	http://www.telescopehouse.co.uk
Astronomics	http://www.astronomics.com
Oceanside Photo and Telescope	http://www.optcorp.com/default.aspx

Autostar Software:

Autostar Update	http://www.meade.com/support/auto.html

Appendix D

Accessories:

Astro Engineering LXD Accessories	http://www.telescopehouse.co.uk
Baader Planetarium – Astro Solar Film	http://www.baader-planetarium.com
Berlebach Tripods	http://www.berlebach.de
BlueStar Bluetooth Adaptors	http://www.starrynight.com
Celestron	http://www.celestron.com
Denkmeier Binoviewers	http://deepskybinoviewer.com
Hutech – IDAS light pollution filter	http://www.hutech.com
JMI Telescope Accessories	http://www.jimsmobile.com
Kendrick Dew Management Systems	http://www.kendrickastro.com/astro/rings.html
LensPen optical cleaner	http://www.lenspen.com
Lumicon Filters	http://www.lumicon.com
Orion Filters and Red Dot Finders	http://www.telescope-service.com/OrionUSA/start/orionUSstart.html
StarGPS	http://www.stargps.ca
Tele Vue	http://www.televue.com
Telrad Finders	http://www.company7.com/telrad/products/telrad.html
Thousand Oaks Solar Filters	http://www.thousandoaksoptical.com
Vixen	http://www.vixenamerica.com/StartPage
William Optics	http://www.william-optics.com

Astrophotography and Imaging:

Adirondack Astronomy - StellaCam astronomical imager	http://www.astrovid.com
Meade LPI and DSI cameras	http://www.meade.com
Mintron Enterprises Co. Ltd – manufacturer of Mintron CCD video cameras	http://www.mintron.com
QuickCam and Unconventional Imaging Astronomy Group	http://www.qcuiag.co.uk
SAC Imaging	http://www.sac-imaging.com/main.html
Santa Barbara Instrument Group CCD Cameras	http://www.sbig.com
Starlight Express (UK)	http://www.starlight-xpress.co.uk

Appendix D

Catalogs:

Caldwell	http://www.astroleague.org/al/obsclubs/caldwell/cldwlist.html
Messier	http://seds.lpl.arizona.edu/messier/ Messier.html
NGC and IC	http://www.seds.org/~spider/ngc/ngc.html
SAO	http://www.to.astro.it/astrometry/Astrometry/DIRA2/DIRA2_doc/SAO/SAO.HTML

Online Maps:

Google Earth	http://earth.google.com
Google Maps	http://maps.google.com
MSN Map	http://maps.msn.com
Multimap	http://www.mulitimap.com

Astronomical Associations:

American Astronomical Society	http://www.aas.org
Astronomical Society of the Pacific	http://www.astrosociety.org/index.html
British Astronomical Association (UK)	http://www.britastro.org/baa
Federation of Astronomical Societies (UK)	http://www.fedastro.org.uk

Magazines:

Astronomy	http://www.astronomy.com
Astronomy Now (UK)	http://www.astronomynow.com
Sky and Telescope	http://skytonight.com
Sky at Night Magazine (UK)	http://www.skyatnightmagazine.com

Software:

Adobe Photoshop	http://www.adobe.com/products/photoshop/index.html
AstroStack	http://www.astrostack.com
AstroVideo	http://www.coaa.co.uk/astrovideo.htm
IRIS	http://www.astrosurf.com/buil/us/iris/iris.htm
K3CCD Tools	http://www.pk3.org/Astro
Mars Profiler	http://skytonight.com
Maxim DL	http://www.cyanogens.com
Registax	http://registax.astronomy.net
Sky Map Pro	http://www.skymap.com
Starry Night	http://www.starrynight.com
The Sky 6	http://www.bisque.com

APPENDIX E

Astronomical Image Information

E.1 David Kolb Images

Location: Kansas, USA
Telescope: Meade LXD55 SC-8
All figures listed below, had final Processing done in Paint Shop Pro (color balance, unsharp mask, histogram adjustments). Noise filtering done with Neat Image Pro.

Table E.1. David Kolb Image Capture Description

Figure	Description

8.8	**The Planet Mars:**
	Date: 26 October 2005 at 7:10:44 UT
	1700 Frames stacked with Registax.
	Capture Software: K3CCDTools
	Camera: ToUcam Pro
	   Settings: 1/25 sec exposure
	   Brightness: 46% Contrast: 51%
	   Saturation: 100% Gamma: 0%
	   Gain: 38% Frame Rate: 10 fps
	   Video Length: 4 minutes Barlow: 5X Powermate

8.9	**The Planet Jupiter:** Date: 16 May 2005 at 3:54 UT Camera: ToUcam Pro 1000 frames stacked with Registax

Table E.1. (Continued)

Figure	Description	
	Capture Software: K3CCDTools	
	Settings: 1/25 sec exposure	
	Frame Rate: 5 fps	Video Length: 2 minutes
	Barlow: 3X	
8.10	**The Planet Saturn:**	
	Date: 9 April 2006 at 2:51:44 UT	
	Camera: ToUcam Pro	
	1600 frames stacked with Registax	
	Capture Software: K3CCDTools	
	Settings: 1/25 sec exposure	
	Brightness: 72%	Contrast: 51%
	Saturation: 100%	Gamma: 48%
	Gain: 46%	Frame Rate: 10 fps
	Video Length: 3 minutes	Barlow: 3X
8.11	**The Planet Uranus:**	
	Date: 6 August 2005 at 8:14:14 UT	
	600 Frames stacked in Registax.	
	Capture Software: K3CCDTools	
	Settings: 1/25 second exposure	
	Brightness: 50%	Contrast: 51%
	Saturation: 50%	Gamma: 100%
	Gain: 50%	Frame Rate: 10 fps
	Video Length: 2 minutes	Barlow: 3X

E.2 George Tarsoudis Images

Location: Alexandroupolis, Greece
Telescope: LXD75 SC-8

Table E.2. George Tarsoudis Image Capture Description

Figure	Description
5.1	**Star Trails near North Celestial Pole:**
	173 images at ISO 400
	About 3 hours exposures with stop steps of 30 second
	Camera EoS 350d 18–55 mm lens on a photographic tripod.
	Image stacked with program Startrails and final image processing in the Photoshop. One dark frame used.
8.4	**The Sun through a Glass Filter:**
	Prime focus with EoS 350d
	Focal reducer f6.3
	Photography protocol:
	Tv (Shutter Speed) 1/2500
	Speed: ISO 100
	Digital image processing with PS

Appendix E

Table E.2. (*Continued*)

Figure	Description

8.5	**Ten Interesting Moon Features:**
	Prime focus with EoS 350d
	Date/Time: June, 12 2006 00:45:59 LT
	Settings: ISO Speed 100
	  Shooting Mode Manual Exposure Tv (Shutter Speed)
	  1/320 Av (Aperture Value) 0.0
	  Metering Mode Partial Metering White Balance
	  Mode Auto AF Mode Manual focusing

E.3 Dieter Wolf Images

Location: Munich, Germany
Telescope: Meade LXD55 SN-10
Other information: Unguided and PHILIPS ToUCam SC1-modified
Single shots acquired with 'K3CCD Tools', stacked with 'Giotto'. Photo finalized with 'PaintShop pro 7'.

Table E.3. Dieter Wolf Image Capture Description

Figure	Description
8.12	**M27 Dumbbell Nebula in Vulpecula:** Date: Three image composite taken in 2005 and July 12 2006 Frame exposure time: 15–30 seconds Total exposure time: 40 minutes.
8.13	**M51 Whirlpool Galaxy in Canes Venatici:** Two image composite Date: 11 May 2006 and May 22 2006 Exposure time: 50×20 and 78×25 seconds
8.14	**M82 Galaxy in Ursa Major:** Date: 4 April 2005 Exposure time: $80 \times 30s$
8.15	**Central Region of M42 with Trapezium:** Date: 28 Jan 2006 Exposure time: 75 frames between 1 and 30s each
8.16	**M57 Ring Nebula in Lyra:** Date: 12 July 2006 Exposure time: 75 frames between 6 and 12 seconds each
8.17	**M3 (NGC 5272) in Canes Venatici:** Date: 22 May 2006 Exposure time: 130 single frames with exposure times between 1 and 20 seconds each

E.4 Author's Images

Location: London, UK
Telescope: Meade LXD55 AR-6 (OTA), LXD75 SC-8

Table E.4. Author's Image Capture Description

Figure	Description
8.1	**Solar Projection through an AR-6 Refractor:** Date: 16 June 2006 Meade 4000 26 mm Eyepiece Taken with Nikon Coolpix 5200 Digital Camera
8.6	**Mercury Transit May 2003:** Date/Time: 7 May 2003 Meade LXD AR-6 Creative Web Blaster web camera at prime focus Home Made Baader Planetarium Solar Filter with yellow filter Processed in Paint Shop Pro (unsharp mask)

8.7	**Venus Transit Image June 2004:**
	Date: 8 June 2004
	Meade LXD AR-6
	Olympus OM1 Camera at prime focus
	Settings:
	  Kodak ISO 200
	  Exposure time: 1/125th second
	Home Made Baader Planetarium Solar Filter with yellow filter
	Processed in Paint Shop Pro

10.1	**Wide Field Image of Cassiopeia Region of the Milky Way OM1 piggybacked on an LXD75 SC-8:**
	Location: Kelling Heath Skycamp Norfolk, UK
	Date: September 2005
	Settings:
	  Kodak Max ISO 400
	  50 mm Lens
	Exposure time: 5 minutes

10.5	**The Moon Taken through a Nikon Coolpix Digital Camera:**
	Date: 9 June 2006
	Afocal Projection with Meade LXD75 SC-8
	Settings:
	  Macro facility set to on
	  ISO 400
	  Optiplex 40 mm

Index

American Astronomical Society, 22, 245
Antoniadi Scale, 19
Astronomical Reference Material, 20
Astrophotography
   afocal projection, 170
   digital image processing, 177
   exposure settings, 172
   eyepiece projection, 170
   focusing tips, 177
   piggyback, 168
   piggyback camera mounts, 206
   prime focus, 169
   video cameras, 176
   webcams, 173
Astroshed, 194
Autostar
   #505 cable set, 158
   description, 50
   firmware upgrade, 169
   menu options list, 239
   mode status display, 240
   object database, 130
   placeholder, 207
   RS-232 serial port, 159
   slew rates, 110
   AUX port, 205
Autostar Options
   Asteroids, 143
   Az/RA percent, 74
   Battery Alarm, 212
   Brightness Adj., 212
   Browse, 153
   Caldwell Objects, 148, 234
   Calibrate Motor, 73, 223
   Clone, 162
   Comets, 144
   Constellations, 145, 229
   Contrast Adj., 212
   Date, 109
   Deep Sky, 147
   Double Stars, 145
   Download, 163
   Easy alignment, 99
   Event, 135
   Eyepiece Calc. Field of View, 119
   Eyepiece Calc. Magnification, 119
   Eyepiece Calc. Suggest, 155
   Exotic objects, 152
   Galaxies, 149
   Getting Started, 108
   Goto, 98
   Guided Tour, 129
   High Precision, 125
   IC Objects, 148
   Identify, 153
   Jupiter, 140
   Landmarks, 138
   Lunar Eclipses, 135
   LXD55/75 Adjust, 71
   Mars, 139
   Max Elevation, 116
   Mercury, 136
   Messier Objects, 125, 148, 231
   Meteor Showers, 144, 237
   Min of Algol, 146
   Moon, 135
   Moonrise, Moonset, Moon Phases, 135
   Named Objects, 148
   Named Stars, 145, 215
   Nearby Stars, 147
   Nearby Stars with Planets, 147
   Nebulas, 151
   Neptune, 142
   NGC Objects, 148
   One Star alignment, 99
   Park Scope, 111
   PEC Erase, 79
   PEC On/Off, 80
   PEC Train, 79
   PEC Update, 80
   Planetary Nebulas, 151
   Pluto, 142
   Quiet Slew, 112
   SAO Catalog, 146
   Satellite, 139
   Saturn, 141

Index

Autostar Options (*cont.*)
  Site, 109
  Sleep Scope, 112
  Smart Drive, 79
  Solar Eclipses, 135
  Solar System, 130
  Spiral searching, 125
  Star, 145
  Star Clusters, 152
  Statistics, 162
  Sun Warning, 108
  Sunrise, Sunset, 135
  Sync function, 127
  Targets, 103, 138
  Three Star alignment, 100
  Time, 109
  Tracking Rates, 121
  Train Drive, 77
  Two Star alignment, 100
  Uranus, 142
  User Objects, 155
  Variable Stars, 146
  Venus, 137

Barlow Lens, 197
Binocular viewers, 203
Bluetooth connectivity, 204
British Astronomical Association, 20, 245

Camera Film
  print processing, 167
  reciprocity failure, 137
  sensitivity, 166
Catadioptrics, 27
Circumpolar, 9
Collimation, 179
    AR refractor collimation, 182
    laser collimators, 182
    N6 reflector collimation, 187
    SCT collimation, 185
    SNT collimation, 183
    star testing, 180
    tools, 182
Constellations
  Big Dipper, 14, 88
  charts, 218–227
  list, 229
  Orion, 15
  recognizing, 13
Counterweights, 49
Cradle
  adjustment screws, 70
  lock knob, 49
  rings AR,N-6,SNT, 61
  rings SCT, 63

Dec Axis
  alignment with OTA, 68
  balancing, 61–62
  general maintenance, 192
Declination, 10
Dew, 26
Dew Shield, 26, 47, 205

Enhanced Multi Coatings, 47
Equipment
  what to take out, 107
Eyepieces, 195
  cleaning, 190
  erecting prism, 197
  kellner, 196
  orthoscopic, 196
  plössls, 196
  zoom, 196

Filters,
  color, 200
  false color correction, 201
  filter wheels, 202
  light pollution, 201
  nebula, 202
  use for imaging 175
  solar, 132
  variable polarization, 202
Finderscopes, 31, 47, 199
  alignment, 82
First point of Aries, 9
First point of Libra, 10
Focal Reducer, 198
Focusers, 30, 47
  adjustment, 84
  motorized, 203

Goto
  setup, 98
  success or failure, 101
GPS, 89, 109
  Autostar add-on, 203

Hartmann Mask, 177
Hypertuning, 192

Isopropyl alcohol, 189

LensPen, 191, 244
Lunar Planetary Imager, 174
LXD AR Refractor, 41
  Specifications, 42
LXD Motors, 72
  backlash, 73
  calibration, 73

Index

cogs, 76
physical alignment, 74
smart drive, 79
training, 77
troubleshooting, 211
LXD Mount
    arrow markers, 67, 213
    azimuth and latitude adjusters, 49
    features, 48
    latitude setup, 91
LXD Newtonian, 41
    specifications, 43
LXD Schmidt-Cassegrain Telescope, 45
    specifications, 46
LXD Schmidt-Newtonian Telescope, 44
    specifications, 45
LXD tripods, 51, 203
    leveling, 81
    setup, 90
    vibration suppression pads, 203

Magnitude,
    absolute, 17
    apparent, 16
    limiting, 17, 117
Maxim DL, 177
Moon, 135
    features, 136

Newtonian telescope, 26
    disadvantage, 27
    Schematic, 26

Observing
    across the meridian, 116
    averted vision, 120
    dark adaptation, 119
    finding the sun, 133
    in comfort, 115
    without use of Goto, 102
    record, 21
    the sun safely, 130
Octans, 92, 101

PC Serial Port Settings, 158
Polar Alignment,
    drift method, 97
    manual two star method, 94
    northern hemisphere, 87
    southern hemisphere, 102
Polar Home Position, 67
Polar viewfinder, 48
    alignment, 63
    internal view, 63, 92

Polaris, 14, 87
    locating in polar viewfinder, 91
Power supplies, 204
Precession, 9

RA Axis
    alignment with OTA, 68
Refractor, 24
    disadvantages, 26
    schematic, 25
Registax, 175, 176
Right Ascension, 9

Schmidt telescope, 27
Schmidt-Cassegrain, 28
    advantage, 29
    disadvantage, 29
    Schematic, 29
Schmidt-Newtonian, 27
    Schematic, 28
Seeing conditions, 18
Setting Circles, 33
    digital, 126
    Manual, 49
    RA calendar dial, 92
Sidereal day, 7
Single Lens Reflex Camera,
    digital 175
    traditional, 165
Solar day, 7
Solar year, 8
Star
    classification, 18
    colors, 18
Star Diagonal, 198
Supercharging, 192

Telescope
    assembling, 56
    balancing, 60
    carry handle, 206
    cleaning, 189
    covers and cases, 208
    exit pupil, 119
    general maintenance, 191
    handling the AR OTA, 58
    handling the mount, 58
    handling the SCT OTA, 60
    handling the SNT, 59
    handling the tripod, 57
    magnification, 118
    packing up, 111
    resolving power, 118
    setup, 108

Index

Telescope (*cont.*)
　slewing, 109
　storing, 193
　tracking problems, 73, 211
Telescope Mounts, 31
　altazimuth, 32
　equatorial, 33
　movement and vibration stability
　　tests, 32

Telescope Stand, 34
　Pillar Stand, 37
　Tripod Stand, 35
　　disadvantages, 36

UHTC, 47
USB Connection, 159

Zenith, 9

Other Titles in this Series

(continued from p. ii)

A Buyer's and User's Guide to Astronomical Telescopes and Binoculars
Jim Mullaney

Care of Astronomical Telescopes and Accessories
A Manual for the Astronomical Observer and Amateur Telescope Maker
M. Barlow Pepin

The Deep-Sky Observer's Year
A Guide to Observing Deep-Sky Objects Throughout the Year
Grant Privett and Paul Parsons

Software and Data for Practical Astronomers
The Best of the Internet
David Ratledge

Digital Astrophotography: The State of the Art
David Ratledge (Ed.)

Spectroscopy: The Key to the Stars
Keith Robinson

CCD Astrophotography: High-Quality Imaging from the Suburbs
Adam Stuart

The NexStar User's Guide
Michael Swanson

Astronomy with Small Telescopes
Up to 5-inch, 125 mm
Stephen F. Tonkin (Ed.)

AstroFAQs
Questions Amateur Astronomers Frequently Ask
Stephen F. Tonkin

Binocular Astronomy
Stephen F. Tonkin

Practical Amateur Spectroscopy
Stephen F. Tonkin (Ed.)

Amateur Telescope Making
Stephen F. Tonkin (Ed.)

Using the Meade ETX
100 Objects You Can Really See with the Mighty ETX
Mike Weasner

Observing the Moon
Peter T. Wlasuk

Printed in Singapore